CONTENTS

PREFACE TO THE PAPERBACK EDITION

The earth is a dynamic planet. And, like the planet, the field of earth sciences is constantly evolving. In the two years since the hardcover edition of this book went to press, old ideas have been refined and many new discoveries have been made about the earth and our neighboring planets. Some of these have come from field-work—for example, the discovery of a new type of mammalian fossil in Australia, which is prompting reexamination of the entire timescale of mammalian evolution—and some have come from the ever higher resolution and precision with which the earth is being probed by new, more sophisticated instruments.

In a short preface it is impossible to do justice to the many important findings that have impacted the earth sciences over the past few years. But I would like to point out a few that are directly relevant to themes discussed in this book. One involves the comet Hale-Bopp, which appeared in April and May of 1997, awing hundreds of millions of people around the globe. It was a milestone in confirming both the role of impacts on our planet and the importance of comets as carriers of molecules that were likely the precursors of life on earth.

Hale-Bopp was a once-in-a-lifetime event, bright enough to be seen even through the light pollution of New York City or London. Its diameter was 30 to 40 km—three or four times the diameter, or thirty to sixty times the volume, of the impactor that was at least partly responsible for the mass extinction at the end of the Cretaceous period. Although Hale-Bopp passed close to us, there was no danger of collision. But other objects of this size *have* crashed into the earth with almost unimaginable consequences. Furthermore, as Hale-Bopp approached the sun and warmed up, it spewed off huge amounts of volatile materials that were observed by astronomers—including water and a long list of molecules such as formaldehyde and methyl cyanide. Such compounds have been observed in molecular clouds far beyond our solar system; Hale-Bopp showed that comets could deliver them to the inner planets, strengthening the argument that organic molecules from space were important for the development of life on earth.

Other findings from the past few years have also led to significant advances in understanding in the earth sciences:

- "Chemical fossils" discovered in ancient sedimentary rocks have pushed back the earliest evidence for life on earth by almost 400 million years, to at least 3.85 billion years ago.
- Studies of seismic waves passing through the earth's metallic core indicate that the inner (solid) part of the core is spinning faster than the earth's surface. This has important implications for the magnetic field.
- Examination of written records from Japanese coastal villages, together with studies of tree rings in Washington and Oregon, indicate that a gigantic earthquake struck the Pacific Northwest subduction zone in the year 1700, sending a huge tsunami across the Pacific. This knowledge raises concern about seismic safety in this now heavily populated region.
- The precise timing of earth's largest mass extinction event, at the end of the Paleozoic era, has been determined by careful study of fossil populations coupled with high-resolution dating of zircon crystals from rocks on either side of the boundary. The extinctions were very abrupt, lasting only a few hundred thousand years; this evidence puts strong constraints on the possible extinction mechanisms.

Although we have learned much about the earth and its history, we have really only scratched the surface. The search for knowledge about how our planet works will occupy us for a very long time to come.

ACKNOWLEDGMENTS

A book like this owes much to many people. I am especially grateful to my colleagues—associates, students, and friends, past and present—at the Scripps Institution of Oceanography, from whom, over the years, I have learned a great deal. They have provided Scripps with an atmosphere in which exploration and learning take place every day. Without that ambiance, I'm not sure this book would ever have been started, or my enthusiasm for it sustained.

At home, Sheila Macdougall endured many dull evenings and weekends without much complaint as this was being written, for which I am ever grateful. Christopher and Katherine were guinea pigs for the raw manuscript, and declared that it could both be read and made sense of by nonscientists, and didn't immediately put them to sleep.

Al Levinson, encouraging throughout, read parts of the manuscript and made many useful suggestions. Rick Balkin was a great help and encouragement at many steps along the way, for which I am very grateful. Guy Tapper drew or reproduced all of the illustrations with professionalism and within a short time frame, and his efforts add considerably to the book. At Wiley, Emily Loose enthusiastically endorsed this project, and shepherded the manuscript over various hurdles, ensuring its timely appearance. To all of you, my thanks.

1

READING
THE ROCKS

IN THE MIDDLE of the seventeenth century, James Ussher, a widely respected scholar and prelate of the Anglican Church in Ireland and England, calculated that the earth was created in the year 4004 B.C. He arrived at this conclusion by careful study and literal interpretation of the genealogies in the Bible. In the time-honored tradition of research, other scholars of his day—themselves not having devised any alternative methods for finding the earth's age—checked Ussher's calculations. He got the year right, they declared, but it was possible to be more precise: The earth was created at 9 A.M. on October 23, 4004 B.C.!

Today some university geology departments, with tongue-in-cheek deference to Ussher, hold birthday celebrations for the earth on October 23. But in fact, the earth is nearly a million times older than the Reverend Ussher calculated. Its real age is 4.5 *billion* years, although it was more than a century after Ussher's writings were published that geologists began to realize the true immensity of geologic time.

Our planet is thus incredibly old by human standards: Four and a half billion years is a stretch of time that has no relevance to human experience. Geologic timescales are so vast that only by analogy can we glean some understanding of the seemingly infinite stretch of time between us and the creation of Earth. One such analogy presents the history of the earth as a three-hour movie. We—as a species, not you and I personally—would make a cameo appearance in the last second or so. This book, like the three-hour movie, is a very abbreviated journey through that history, from the formation of the solar system to the present. It is arranged chronologically, with occasional diversions to discuss topics that are important for understanding the history. But the reader should be warned that it touches only some of the highlights. One could easily consume several lifetimes mastering all of the details of the earth's fascinating past.

For most of us, the natural landscape has a degree of permanence. Barring calamities such as volcanic eruptions or large earthquakes, the geologic panorama really doesn't change perceptibly in a human lifetime. But the earth in *its* lifetime has witnessed extraordinary transformations. Over the billions of years of its existence, our planet has endured global catastrophes on a scale unequaled in human experience, has seen the rise and fall of countless species that no longer grace the earth, and has watched entire ocean basins and mountain ranges form and then disappear. How do we know such things? Some of our understanding comes from laboratory experiments and from mathematical simulations of geologic processes, or even from intelligent speculation, but much of it comes from the rocks. Rocks are the recorders of the earth's history, and hold the clues to its past. Deciphering them is not always easy, and although a great deal has already been learned, there is much more to discover. This book is meant to whet your appetite for such knowledge, because there are few things more satisfying than understanding the origins of one's physical surroundings, and, perhaps, one's place in that world.

The earth sciences, like other disciplines, are replete with jargon. Partly this is because rocks, minerals, fossils, and landforms need names if they are to be discussed intelligently. It also has to do with the vast length of geologic time: Geologists have subdivided the earth's history into units of time, and have given them names that are unfamiliar to most nongeologists. These names are usually based on a specific geographic locality where rocks of a particular time period are especially prominent. I have tried to keep geologic jargon to a minimum in this book, but unfamiliar words will recur, some of them frequently. There is a short glossary at the end for

reference. Figure 1.1 should also help you get a grasp of the geologic timescale. The timescale is the bane of students taking introductory classes in geology, but most succumb and learn the names
of the eras, periods, and even more detailed subdivisions after
being reminded that there are some things you just learn—such as
the months of the year, or the statistics of your favorite baseball
team. Soon it becomes second nature.

The boundaries between the eras, periods, and epochs of the
geologic timescale were originally defined mainly on the basis of
fossils, which are part of the record of the rocks. Throughout the
earth's history, species and families have arisen, prevailed for a
time, and then disappeared. But at times, for reasons not wholly
understood, rapid and wholesale destruction of large fractions of
the plant and animal kingdoms has occurred. Usually, after these
crises, there was a rapid proliferation of new and sometimes quite
different species. Such abrupt changes in floral and faunal assemblages are reflected in the fossil record. It is only quite recently that
geologists have begun to examine these mass extinctions in terms
of periodic catastrophes such as the collision of comets or asteroids with the earth, or dramatic changes in the global climate.
Nevertheless, although the interpretations may change, the records
of these events have always been present in the rocks for all to see,
and they provided a logical way for early workers to compartmen-
talize geologic history. Boundaries were placed where the nature
of the fossil record changed drastically. A simple version of the
geologic timescale is shown in Figure 1.1. You will probably want
to refer to it frequently as you read this book.

The connection between the timescale and rocks may not, at
first, be obvious. But the picture becomes clearer when one con-
siders how sedimentary rocks, which were used to define the
timescale in the first place, are formed. Sediments accumulate at
the earth's surface piece by piece, sometimes atom by atom, usually
in water. They are the result of erosion and weathering on land,
their constituents carried to lakes or to the sea by streams. Most
sediments begin as unconsolidated material such as mud or sand,
and, through a variety of processes, harden into coherent rock.
Sediments engulf and preserve shells, skeletons, leaves, feathers,
and other parts of plants and animals, and thus provide a record of
biological evolution. A single outcrop of these rocks may span

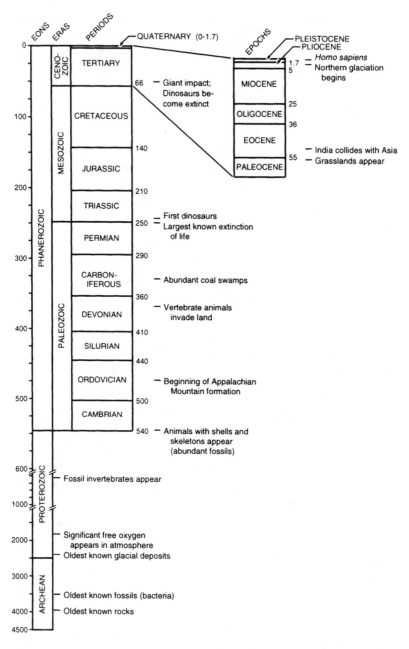

FIGURE 1.1   *The geologic timescale. Times are shown in millions of years, and a few important events in the earth's history are indicated. Note the two changes in scale in the Proterozoic section of this timescale.*

thousands or even millions of years of uninterrupted deposition—and the oldest layer is always at the bottom, the youngest at the top. Much of the timescale shown in Figure 1.1 has been built up by piecing together, from many parts of the world, portions of this record that, based on their fossil contents, are overlapping. But it should be recognized that both sedimentation and the preservation of fossils are capricious processes. Furthermore, when sea level drops, or piles of sediment are uplifted, erosion occurs, wiping out a part of the record. Consequently there are many gaps. This was a serious problem for Darwin, who had to explain why the fossil record didn't show every evolutionary step in detail. An entire chapter of *The Origin of Species* is devoted to this issue under the title "On the Imperfection of the Geological Record."

However, although they provide the best continuous historical record, sedimentary rocks are not the only materials of interest to geologists. Igneous and metamorphic rocks also contain information about their origin and history, although of a different type. In contrast to sediments, igneous rocks originate in the earth's interior by melting, and crystallize to their present state when the molten magma—the term used by geologists for liquid rock—cools on or near the surface. Familiar examples are the pink granite that graces the facades of banks and other buildings, or the dark-colored basalt that forms from lava pouring out of Hawaii's Kilauea volcano. The chemical compositions of such rocks contain clues about the geologic settings in which they were created. While this may not seem very earthshaking for young materials—we know that Hawaii is a volcano in the middle of the Pacific without measuring the chemical composition of its lavas—it is critical for understanding ancient rocks, because it allows us to reconstruct the physical world of the past.

Metamorphic rocks are different altogether. Originally sedimentary or igneous material, they have been changed significantly, usually by deep burial and heating that transform their mineral makeup. Their very existence is a sign of the changeability of the earth over long time periods. Metamorphic rocks that we walk over or climb on without a second thought may have begun life in the distant past as grains of weathering debris, accumulating layer upon layer in the seas along the edge of an ancient continent. The metamorphic minerals they now contain, however, bear mute witness to another, less passive, stage in their history: burial, perhaps to depths of twenty kilometers or more, and strong heating. This often occurs during a mountain-building episode, and we know that such metamorphic rocks exist deep within the cores of the Andes and the Himalayas today. But how can we ever expect to see such materials at the earth's surface? The answer lies in the fact that even spectacular mountain ranges are ephemeral by the standards of geologic time. Victims of slow but steady erosion and uplift, they are gradually worn away. Our deeply buried sediment, now a metamorphic rock, is eventually exposed again at the surface by this process. Such cycles are a natural part of the earth's operation, and although they occur on timescales far too extended for direct human observation, they leave their record in the rocks.

Not long ago, even geologists didn't really understand why there are volcanoes in Japan, or why the Ural Mountains formed in central Russia. The theory of plate tectonics changed all that. Suddenly geology, like most other disciplines, had an underpinning, a foundation through which many seemingly disparate observations could be understood. In this view, the earth's surface consists of a series of large rigid plates, roughly 100 kilometers thick, that move about, relative to one another. In some places the plates are splitting apart, growing in size by addition of material at the diverging boundary; in other places plates collide, usually with the result that one plunges beneath the other into the earth's interior. In yet other locations the gigantic plates simply slide by one another, grinding up the earth's crust in the process, as is happening along the San Andreas Fault in California. Nearly all geologic action occurs at the plate boundaries. If you were to plot the locations of all earthquakes that have occurred over the past decade on a map of the world, they would neatly outline the tectonic plates. Most of the world's volcanic activity also takes place along these boundaries.

The plate tectonic map of the world is a giant jigsaw puzzle, each piece a plate, with the important difference that the pieces are moving and, slowly but surely, their shapes keep changing. Make the same map fifty million years from now, and Los Angeles would be on an island somewhere off central British Columbia, and Australia scrunched against the islands of Indonesia. New York would be farther from London but closer to Tokyo than it is now, because the Atlantic would have widened at the expense of the Pacific.

Contrary to some popular accounts, the plates don't move about on a liquid interior like ice floating in water. Instead, they move by a kind of plastic flow at their base. The earth's interior is solid, but it is also hot, allowing it to deform and flow by slow movement over long time periods, much in the way glacial ice flows. By contrast, the surface plates are cold and quite rigid. Their physical properties disconnect them from the underlying, convecting, interior of the earth.

Convection in the interior is actually the principal mechanism by which the earth loses heat. The rocks that make up the earth's mantle (see Figure 1.2) are such good insulators that it would take many billions of years to pass heat from the interior to the surface by conduction alone. However, convection in the mantle physically moves hot material toward the surface, and a balancing return flow transports cooler material into the interior. It is probable that this convective circulation in the mantle is at least partly responsible for the movement of the surface plates.

Although the earth's interior is largely solid, part of the core (see Figure 1.2)—the very dense, central part of the earth that makes up about one-third of its mass—appears to be liquid. More will be said later about the core, but for the moment it is sufficient to note that it is composed mostly of metallic iron, and that it is because of convection in its liquid outer part that the earth has a magnetic field. We know this in spite of the fact that no one has ever obtained samples of the core. Jules Verne's imagination aside, no human has ever been more than a few kilometers below the surface, and even the deepest drill holes are less than 10 kilometers deep. In contrast, the outer boundary of the core lies at a depth of 2,900 kilometers, and it extends downward from there to the earth's center at approximately 6,370 kilometers.

Lacking direct information about the interior, we have to be satisfied with evidence from remote sensing. By far the most useful such evidence has come from studies of the way in which earthquake waves are transmitted through the earth. Large earthquakes obviously release huge amounts of energy, and the resulting vibrations travel through the earth in the form of waves. They can be recorded at remote locations, much in the way that the vibrations caused by striking one end of a table with a hammer can be felt at the other end. The wiggles on a seismogram are the instrument's response to actual vibrations of the earth. The details of interpretation of the seismic records are complex. However, the net result of years of recording earthquake signals at many stations around the globe is a body of data that makes it possible to determine the average speed of waves passing through different parts of the interior. Because these seismic wave velocities are directly related to the density of the material in which they are traveling, it has been possible for geophysicists to calculate the density, and from this to deduce what minerals are present, at different locations in the earth. The data show that the earth has a layered structure (see Figure 1.2), and that the major subdivisions have quite different densities and chemical compositions. Although Figure 1.2 is a simplified picture, it indicates that the earth is chemically differenti-

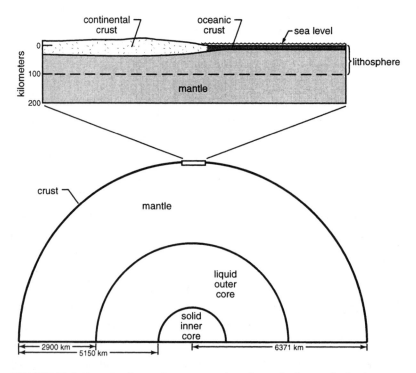

FIGURE 1.2   *A schematic cross section through the earth showing its layered structure. A blowup of the outer part illustrates that the continental and oceanic crust differ in thickness, and that both are part of the lithosphere, the rigid outer skin of the earth that forms the plates of plate tectonics.*

ated on a large scale. This is an observation with important ramifications for the early history of our planet, because most scientists believe that these now-separated components of the earth were intimately mixed together in a more or less homogeneous mass when the earth formed. As far as can be discerned from the available evidence, the other Earth-like planets (Mercury, Venus, and Mars), as well as the moon, have also undergone global chemical differentiation.

Most of this book deals with processes taking place on or within the earth's crust—the upper crust at that. A glance at Figure 1.2 shows that, in terms of mass or volume, the crust is quite insignificant compared with the other subdivisions of the planet. It is a thin skin on the earth, only 5 or 6 kilometers thick under the oceans, and 30 to 40 kilometers thick under the continents. If the earth could be shrunk to the size of an apple, the thicker parts of the crust would be roughly the thickness of the apple skin. But in spite of this, it is in the crust that mineral deposits reside, it is where life originated, and it is where we make our home. It is the best-known part of the earth, because it can be explored, analyzed, and measured. It has been formed over geologic time by melting of the planet's interior and upward transport of buoyant liquids to the surface.

The boundary between the crust and the underlying mantle is marked by a sharp increase in the velocity of seismic waves, reflecting a change to the denser rock types of the earth's interior. The rocks of the mantle are richer in iron and magnesium, and poorer in lighter elements such as aluminum, than those of the crust. This is known both from seismic studies and from actual samples. How is it possible to get samples from the mantle when even the deepest drill holes don't penetrate through the entire crust? It turns out that nature has done the sampling for us: There are a few places on the earth where volcanic lavas, originating in the mantle, have actually prized off solid fragments of the surrounding rocks and carried them to the surface. One consequence of this process is that we can wear diamond jewelry. Diamonds are a form of carbon, which is also the main constituent of charcoal—not exactly a popular material for personal adornment. However, at the high pressures that exist in the mantle ordinary carbon transforms to diamond. The pressures required correspond to depths in the earth of about 200 kilometers; the diamonds of South Africa and elsewhere have been carried to the surface in volcanic magmas that formed at least this deep. Of course, finding these gems from the mantle is not to say that the interior is composed of diamonds—the diamonds themselves are rare, and it is the solid rock fragments in which the occasional diamond is found that provide the real clues about the makeup of the mantle.

Figure 1.2 shows that the plates on the earth's surface actually contain both crust and mantle material. Their base is not marked by a change in rock types; rather it is a physical boundary at which the seismic wave velocities decrease considerably. It is generally believed that this is the depth at which mantle rocks are closest to their melting point, and, because of the increased temperature and pressure, behave plastically, allowing the overlying rigid plate to move over the underlying convecting mantle. The rigid outer part of the earth, the part that constitutes the plates of plate tectonics, is known as the lithosphere, after the Greek *lithos* for stone or rock.

The mantle accounts for about two-thirds of the earth's mass, and has been subdivided into an inner and outer part based on the subtleties of seismic wave velocities. It is underlain by the core, which constitutes the remaining third of the mass, and which, as has already been mentioned, is made mostly of iron. There is a very large change in the seismic wave velocity at the boundary between the core and the mantle, reflecting the change from rocky material to metal. Some types of waves cannot be transmitted through liquids, and it is observed that these don't pass through the outer part of the core, indicating that it is liquid. The inner core, however, is solid.

No one knows in detail how the earth was formed. However, it is possible to extrapolate from what we do know, and construct a reasonable scenario. We know that the universe is much older than the earth, and that most of the atoms now constituting the air we breathe, the rocks (or concrete, as the case may be) we walk on daily, and all other parts of the earth were once nuclei in the interiors of stars. Some of the heaviest elements, such as gold, lead, and uranium, were made in gigantic supernova explosions that ended the life of a star and spewed large amounts of matter into interstellar space. We know that eventually the material that is now the earth found itself part of a great cloud of gas and dust, much like those that astronomers observe today in other parts of our galaxy.

For reasons that are not well understood, this cloud began to collapse about 4.6 billion years ago. As it collapsed, the central regions became denser and hotter, just as air compressed in a bicycle pump becomes hotter. At the very center of the collapsing cloud, where temperatures and pressures were extreme, the nuclear reactions that fuel the sun began. Our local star, the sun, contains about 99.9 percent of all matter in the solar system; the planets and asteroids are merely the leftover debris. At least in the inner part of the solar system where the earth resides, the heat was so intense as the sun was forming that any preexisting solid grains were probably vaporized, and most of this leftover debris was in gaseous form. As cooling set in, solid grains began to condense and agglomerate, gradually forming larger bodies. Some grew rapidly, sweeping up everything in their paths as they traveled in orbit around the early sun. Others were destroyed in spectacular collisions of large fragments. The accretion process that formed the earth was violent, and the continual rain of impacting bodies must have heated the early earth considerably. Although the initial mix of materials may have been quite homogeneous on a large scale, the heat of accretion led to melting, and the liquids that were produced separated from the unmelted solids under the influence of gravity. In particular, iron, which melts at slightly lower temperatures than many other earth materials, would have melted early, and, because of its high density, sunk rapidly through the hot earth to form the core.

The large-scale chemical differentiation of the earth into a metal core and an overlying rocky mantle must have occurred near the very beginning of earth history. The formation of the crust is a different story. We know that it, too, formed by melting, but in this case the molten materials, in contrast to the molten iron, were less dense than the surrounding mantle, and rose to the surface. This process still goes on: The lavas erupted from most volcanoes today are the products of melting in the mantle, and constitute new crustal material. The crust, in particular the continental crust, has grown over the history of the earth, although whether its growth has been continuous or episodic, or whether its rate of growth has changed with time, are subjects of debate among earth scientists.

Geology is an old science. Early man practiced it in a rudimentary way in order to locate deposits of rocks such as chert or obsidian that could be flaked to make sharp-edged tools for hunting and scraping. Locating the mineral and energy deposits that provide the materials necessary for modern society is still an important task for geologists. Equally important is the search for a better understanding of how the earth works, without regard for immediate practical significance. Geology, after all, surrounds us every day, although it may be a bit difficult to recognize that fact if you live in the heart of a big city. But a visit to the Grand Canyon or Yosemite is a wholly different experience after learning a little geology. To recognize that much of the beauty of Yosemite, with its steep cascading waterfalls, is the work of towering glaciers that carved the rocks of the Sierra Nevada during a recent ice age, or to understand the comings and goings of the seas that, many millions of years ago, deposited layer after layer of the sediment now exposed in the walls of the Grand Canyon, is for most people an intensely satisfying experience.

To arrive at our current understanding of the earth and its history, geologists have had to be historians, detectives, explorers, engineers, and above all, keen observers. Increasingly, they also need to be biologists, chemists, physicists, and mathematicians, because the study of the earth embraces all of these fields. The search for answers in the earth sciences quite literally leaves no stone unturned.

2

EARLY DAYS

THE OLD TESTAMENT relates that the earth was made in seven days. Most geologists believe that even God couldn't accomplish the task quite that fast. Nevertheless, it must have happened fairly quickly, geologically speaking. Just how quickly is quite important to know, because the pieces of matter that accreted to form the earth carried with them kinetic energy, and as these fragments collided with the growing earth, the kinetic energy was transformed into heat. The amount of heat that was buried in the rapidly growing earth, rather than being radiated away from the surface into space, determined how hot the earth was at the end of its accretion. The faster the accretion process, the more heat retained and the hotter the newborn earth. Undoubtedly the early earth was very hot, although this is a matter about which we don't have a great deal of information. Was the outer part completely molten? Was there a magma ocean on the earth, analogous to the one that many geologists think existed on the moon? Was the whole earth molten?

There are proponents of all of these views, but no really definitive evidence for any one of them. Unfortunately but inescapably, the geologic clues to the earth's history become ever more sparse and difficult to interpret as one probes farther back into geologic time. In the beginning, as noted in the previous chapter, the earth and the other planets of our solar system formed from bits and pieces of matter that were orbiting around the early sun. The earth grew by sweeping up all the matter in its vicinity, and it reached approximately its present size in a matter of some millions of years, probably ten or less. Although we don't know exactly how quickly it accumulated, we do have some clues about the kinds of matter from which it formed. We get this information from studies of meteorites.

 ## METEORITES AND THE EARTH

Meteorites are more common than you might think. The number of specimens in private and public collections is in the thousands and growing. Most of the shooting stars that streak across the night sky are tiny meteorites that are heated to white-hot temperatures by friction, and actually burn up as they speed through the earth's atmosphere. A few survive the journey and land on the surface. Tens of thousands of meteorites, perhaps even more than 100,000, fall on the continents each year, and even more fall into the oceans. Most of these are very small and are never recognized. Those that have been found and collected range from the size of a pea to rarer objects as big as a basketball or even much larger. As the earth becomes more densely populated, a larger fraction of the meteorites that fall is recognized and picked up immediately. A few have even struck cars and houses.

In recent years, one of the most prolific sources of meteorites for scientific study has been the Antarctic. Meteorites that fall on the ice cap are buried by snow and ice, but eventually they are carried toward the sea by the glaciers moving slowly outward from the pole. Their flow path takes them deep below the surface, but in places where the ice encounters buried mountain ranges, it is deflected upward. In such regions, the cold, dry winds of the Antarctic ablate the ice away as rapidly as it arrives at the surface. However, the meteorites it carries remain. Thousands of years of meteorite falls can be concentrated in a small area by this process, and because there are few other rocks around in this sea of ice, they are easy to spot. Geologists from several countries now mount annual expeditions to the Antarctic in the southern summer to scour likely locations by snowmobile and helicopter, in search of treasure troves of meteorites.

In times past, meteorites were sometimes accorded special powers because they came from the heavens and were thought possibly to be sent from the gods. More recently, it has been recognized that they are truly Rosetta stones that carry information about the earliest history of the solar system. Although there are many varieties of meteorites, some appear to be essentially unchanged since their formation 4.5 billion years ago, about the time the earth formed. In fact, they are probably very much like the original material that accreted to make the earth. Next time you visit a natural history museum, spend some time looking at the meteorites. Although they may look much like ordinary rocks, they are not. Instead, they are amazing messengers from the past, with much to tell us about the time when the solar system was just forming.

The chondrites, as the most primitive meteorites are called, are believed to be fragments of material from the asteroid belt, which lies between Mars and Jupiter. They are made up mostly of minerals that are common in rocks on earth, but they also contain metallic iron, which is very rare as a natural substance at the earth's surface. As we learned in the previous chapter, iron melts at a lower temperature than many common minerals. Most of the metallic iron brought to the earth in chondrite-like materials during the accretion process melted, and sank to the center of the planet to form the core.

Because the earth is made up of chemically different regions such as the core, the mantle, and the crust, and because we can adequately sample only the outermost of these, determining the overall chemical composition of the planet has been a difficult task. However, the chondrites can be analyzed in the laboratory. If they are truly representative of the material that accumulated to form the earth, then simply by analyzing them we should be able to determine the chemical composition of the whole earth—a truly amazing prospect. But are they truly representative of the average solar system material that must be the primary constituent of the earth? There is strong evidence that they are. It comes from studies of the sun, which, because it contains nearly all of the mass of the solar system, is by definition average solar system material. By analyzing light emitted from the sun, we have obtained a great deal of information about its chemical composition. Except for a small number of elements, such as the gases, the relative abundances of most elements in the chondrites are almost exactly the same as in the sun, a good indication that these materials have not undergone any significant chemical fractionation. Thus, by combining information obtained from meteorites with knowledge of the earth's interior density gleaned from seismic studies, it has been possible not only to estimate the earth's overall chemical composition, but even to determine the makeup of regions that have never been sampled, such as the deep mantle and the core.

 ## HOW OLD IS OUR PLANET?

It has already been mentioned that the earth is billions of years old. This is the modern view; the ideas of James Ussher, the theologian who calculated from the biblical record that the earth was created in the year 4004 B.C., held sway well into the nineteenth century. Indeed there are those who even today ignore the overwhelming scientific evidence to the contrary and claim that the biblical legends tell the true story of the earth's creation and history.

It was not until the mid 1950s that the now-accepted age of 4.5 billion years for the earth was ascertained. Determining the earth's age precisely is a highly technical matter, but basically it depends upon the fact that naturally occurring radioactive isotopes decay at a constant rate. If the rate is well known, the amount of the decay products that has accumulated in a sample over time can be measured, and the age determined in a fairly straightforward way. Radioactive decay and its use for dating geologic samples will be explored in more detail in Chapter 6, but it is worth noting here that there are several isotopes found in ordinary rocks that can be used for dating. Isotopes of an element share the same chemical properties, but have slightly different nuclear characteristics. Not all isotopes are radioactive, but those that are break down over time to form new isotopes of a completely different element. Two of the better-known elements with radioactive isotopes are thorium and uranium. During radioactive decay, they are transmuted to isotopes of lead. Thus some of the lead in the earth, and indeed in the solar system, did not exist when the earth was formed, but has been created over geologic time by the gradual decay of thorium and uranium.

Because each of the isotopes of thorium and uranium decays to lead at a different rate, samples containing these elements have several independent built-in geologic "clocks" that can be used to determine their age. This also means that the exact mix of lead isotopes in any particular material is usually quite distinctive, depending both on its age and on the amount of thorium and uranium it contains. In the 1950s, Clair Patterson, of the California Institute of Technology in Pasadena, discovered that both meteorites and samples from the earth share common characteristics with regard to their lead isotope contents. Using samples carefully chosen to represent as closely as possible the average lead isotope abundances for the earth, together with a series of samples from the chondrite group of meteorites, Patterson discovered systematic relationships indicating that all of these bodies—the earth and the various chondrites—must have formed from a common ancestral material between 4.5 and 4.6 billion years ago.

Patterson's finding was an extremely important discovery in the annals of geology. It not only established a reliable age for the earth, but it linked the origin of our planet with that of other solar system materials. An earlier observer, the remarkable eighteenth-century Scottish gentleman-geologist James Hutton, had said about earth history that he found "no vestige of a beginning, no prospect of an end." In spite of Hutton's lyrical prose, however, Patterson's work firmly established the timing of the beginning. And although there have been great technical advances in Patterson's field of isotope measurement since the 1950s, his basic conclusions remain unchallenged.

The number 4½ billion rolls off the tongue quite easily. Students and professors of geology become quite accustomed to it. But it is an enormous number, much too large to comprehend in terms of human experience. Put in the zeros, and the earth's age is slightly easier to grasp: 4,500,000,000 years. Four and a half billion pennies would make a stack about 6½ *thousand* kilometers high, which is somewhat more than the distance from the surface of the earth to its center.

THE FIRST 600 MILLION YEARS

Although we know when the earth formed, the next chapter of earth history is essentially blank. For almost 600 million years after our planet's creation the record of the rocks fails us. The oldest rock samples yet identified on earth come from the Northwest Territories in Canada. Based on analysis of their lead isotopes, they have an age slightly greater than 3.9 billion years. These rocks have undergone strong metamorphism, and it is difficult to tell much about their origin. However, they are not terribly different from many other typical continental rocks that are much younger. Thus we know that there were at least some fragments of the continental crust in existence 3.9 billion years ago.

The question of when the first continents formed is one that has long intrigued geologists, because it is evident that the continental crust has grown and evolved over geologic time. It is probable that there were small continents around even before the creation of the 3.9-billion-year-old rocks. The clues that lead to this conclusion are rare and minute, and almost as difficult to discover as the proverbial needle in a haystack. But where would one look for such evidence? The answer provides a good example of how geologists often work: by using the present as a window into the past. We know that the products of erosion accumulate on the edges of continents today, and there is no reason to suspect that the situation was any different in the past. Even the earliest continents must have had beaches. There is a chance that if some of those very old sediments have been preserved, they might actually contain mineral grains eroded from even older continents.

Geologists have thus sifted through some of the oldest known sandstones, which were probably originally laid down along the shorelines of ancient continents, in search of mineral grains that are especially resistant to destruction during weathering and transport. One such search hit pay dirt in a 3.6-billion-year-old sandstone from Western Australia. Some of the grains in these rocks are much older than the sandstone itself, and have apparently survived multiple cycles of erosion, deposition, consolidation into solid rock, uplift, and re-erosion. William Compston and his colleagues at the Australian National University in Canberra have found that a few grains of the weathering-resistant mineral zircon from these old sandstones have ages in the range of 4.1 to almost 4.3 billion years.

Zircon crystals are small but common constituents of many igneous rocks. Pick up a handful of beach sand or soil and you may be holding a few zircon grains in your hand, because the weathering and erosion processes that break down their parent rocks have little effect on the inert zircon crystals. Because they are hard and resistant, large, transparent zircon crystals are often sold as gemstones. But the ones that are most useful for geologists are the small zircon grains that are carried long distances in streams or even by wind. These become tracers of the ultimate source for the sedimentary materials in which they now reside.

As their name implies, zircons are rich in the element zirconium. Fortunately, they also incorporate considerable amounts of uranium when they form, and as we have already seen, the radioactive decay of uranium produces isotopes of lead that can be measured to determine the age of the grain. Modern techniques are so sensitive that even the minute amount of lead in a single small zircon grain can be measured precisely, and its age determined. This is how the grains extracted from the Australian sandstone were dated.

Because these ancient zircons are single grains rather than rock fragments, it is difficult to say much about the types of rocks from which they were originally eroded. However, zircons are common in continental rocks such as granite but virtually absent from the ubiquitous basalts of the ocean floor. The inference is that these grains must have come from continental rocks. If this is indeed the case, the existence of continents can be pushed back to almost 4.3 billion years, only a few hundred million years after the formation of the earth. But these early pieces of the earth's crust may have been quite different from the continents we know today, and certainly they must have been much smaller.

Even if the earth's crust began to form very early, there are several possible reasons why none is preserved from approximately the first 600 million years of our planet's existence. One is that for much of this period, the earth was experiencing heavy bombardment from space as the residual material from the accretion process was gathered in by the earth's gravity. A second is that, as we have already noted, the early earth was very hot. Vigorous convection in the hot earth may simply have destroyed much of the early-formed crust. Although a great deal of the heat came from the accretion process itself, some of it must also have come from what was probably the major event of the earth's earliest history: formation of the core.

As the planet heated up during its formation, the metallic iron it contained began to melt, and small pools of molten iron aggregated, eventually reaching quite large sizes. Being much more dense than the surrounding material, they sank toward the center of the earth. The process was aided by the fact that the surrounding minerals, although not molten, were also hot and could flow plastically. It is estimated that a 1-kilometer sphere of molten iron would make the journey from surface to center of the hot early earth in less than a million years.

The process of melting, aggregation, and sinking of iron that led to the formation of the earth's metallic core occurred very early, probably during, and possibly for a short while after, the main phase of accretion. This means that within a few tens of millions of years of its formation, at most, the earth was already a chemically differentiated planet, with metal at its center, and a rocky outer part. This major chemical reorganization from an initially much more homogeneous state has sometimes been called the iron catastrophe, because some analyses have suggested that it was a runaway process, accompanied by a large release of energy, perhaps enough energy to melt the entire earth. In one published description of this event, it was suggested that a large fraction of the metal now in the core agglomerated into a ring or shell of molten material around a cooler, central part of the newly formed earth. As gigantic "droplets" of the molten metal from this shell began to sink toward the center, the change in the distribution of mass within the rotating planet caused huge stresses, violently breaking up still-solid parts of the interior and displacing them with molten iron. Whether this is an accurate depiction of what really happened is unknown. But regardless of the way in which the iron metal made its way to the earth's center, a large amount of energy would have been released, further heating up the earth.

Thus the earliest days of earth history must have been very chaotic, with extensive volcanism and perhaps even a sea of molten rock at the surface. Initially, there was no atmosphere. However, compounds such as water and carbon dioxide, as well as various volatile elements, had been brought to the earth bound up in the accreting material, and they were gradually released from the hot interior as volcanic gases to form an early atmosphere. A steady rain of large and small objects from space smashed through this gaseous envelope and into the surface as accretion gradually wound down. Even several hundred million years after its formation, the earth would have been very unfamiliar and inhospitable for time-machine human visitors. By this time there would probably have been liquid water on the surface, but there was no visible life—no plants, no animals—and the atmosphere was unbreathable because it contained no oxygen. There were no large continents as we know them today, and although there must have been many volcanoes, extensive mountain ranges like the Rockies or the Alps didn't exist.

It may even be that the earth was periodically in a deep freeze during part of its early history, with frozen seas covering most of the surface. This possibility arises from the fact that the sun, if it followed a normal evolutionary path for stars of its size, would have been considerably weaker, with much less energy output than today, for the first part of its life. In spite of the heat from volcanoes and impacts, it is the sun's energy output that ultimately controls the surface temperature of the earth. After the initial hot stage, which could have lasted several hundred million years, the earth's surface cooled, and, with a weak sun, temperatures may have been cold enough for any existing oceans to freeze. In fact, some scientists have pointed out that once the earth was covered with highly reflecting ice and snow, so much of the sun's energy would have been reflected back into space that the surface might never thaw, even with a much stronger sun. They have used this argument, and the fact that the earth today is, in most places, cozily warm, to argue that the early deep freeze never occurred. However, there are other ways to thaw the ice, as we will see in the next chapter.

THE ARCHEAN EON

The first major division of geologic time is the Archean eon (Figure 1.1). It is very long, lasting from the time of earth formation until about 2.5 billion years ago, and thus spanning roughly 44 percent of the earth's history. Of course, the geologic timescale is just a construct of scientists and many things must have occurred during the Archean that, if we only knew about them, might provide a basis for further subdivisions. But in spite of its length, we know very little about the history of the Archean. At least in part this is because only a small fraction of the earth's surface today is made up of rocks that actually formed during this time. We have just seen that there are none at all from approximately the first 600 million years.

Although (or perhaps because) they are scarce, Archean rocks have been the subject of intense study. We know, for example, that they occur, albeit in small amounts, in all major continents. They are sometimes situated near the center and are always surrounded by younger rocks, a configuration that has provided clues about how continents grew. There is evidence from their ages that the continents grew episodically during the eon, but this is uncertain because of the small number of occurrences of Archean rocks, and the possibility that their preservation has been selective. Some fossils have been found in Archean sediments, the remains of ancient, single-celled bacteria. In recent years careful study has revealed that they are more abundant than once thought, but they are still quite rare. Nevertheless, they indicate that life was well established by the middle of the Archean.

We have learned from the Australian zircons that there may have been small continents as early as 4.2 or 4.3 billion years ago. Throughout geologic time, beginning in the Archean and continuing today, continental crust has been produced by melting of the interior and transfer of the molten material to the earth's surface. However, even today the continental part of the earth's crust is a very small fraction of the earth as a whole, as is evident from Figure 1.2, and it also has a very different chemical composition from the rest of the planet. Some of the other planets in our solar system have crusts, but continents as we know them seem to be unique to the earth. In consequence, few, if any, of the diverse mineral deposits that occur on the earth's continents and supply most of the materials necessary for modern civilization can be expected on other planets. Why don't continents exist elsewhere? The answer probably has to do with the presence of liquid water on the earth.

Like salt added to ice, water added to rocks lowers their melting temperature. It also affects the composition of the magma that is produced when melting occurs. On the earth, the process of plate tectonics has the effect of adding water to the already hot interior, triggering melting. Water-rich ocean crust is dragged downward into the mantle at the great ocean trenches, and, as the temperature increases, the water is driven off. It is this process that produces the so-called ring of fire around the Pacific Ocean: The volcanoes of Washington, Chile, Alaska, and Japan all sit above regions where the Pacific Ocean floor is thrust into the interior of the earth, releasing water and initiating melting. The molten material that results is less dense than its surroundings and rises toward the surface, adding new material to the continents from the earth's interior. Although there is considerable debate among geologists about just when the process we call plate tectonics began, the presence of Archean continental crust argues that water was being transferred from the surface to the interior very early in earth history, probably in a manner not too dissimilar to that occurring today.

The Archean eon ended 2.5 billion years ago. Its boundary with the Proterozoic is the only boundary in Figure 1.1 that is not defined primarily on the basis of changes in the fossil assemblage in rocks. Although life was well established by this time, the Archean bacteria were without easily fossilized skeletons or shells, and they are not abundantly preserved. They also did not evolve very rapidly, and therefore are not particularly good time markers. As an index of geologic time, fossils are most useful from the beginning of the Cambrian period onward, when diverse organisms with hard body parts began to flourish. As a result, the age of the Archean-Proterozoic boundary, 2.5 billion years, is in a sense just a convenient number. True, it was chosen based on a general perception, the result of many years of study, that some things in the geologic record change at about this time—for example, the chemical compositions of rocks being formed, and, to the extent it can be discerned, the nature of the few fossil remains that can be identified. But unlike all of the other dividing lines in the geologic timescale, there is no place in the world where you can put your hand on this particular boundary in the rocks.

The oldest Archean rocks that are recognizably sediments are about 3.8 billion years old. These occur in western Greenland, and they confirm that continents and oceans existed by that time, and that erosion and sedimentation were proceeding in ways not radically different from today. But even 800 million years after its birth the earth was still a barren place, and the atmosphere still without oxygen. In spite of this, and although the evidence is only indirect in rocks of this age, life in the form of microbes or one-celled organisms was probably already present. How early life may have arisen in the first place, and how it probably developed, are topics of the next chapter.

3

WONDERFUL LIFE

WONDERFUL LIFE IS the title of a book by paleontologist Stephen Jay Gould of Harvard University about the evolution of life on earth. Gould took his inspiration for the title from the movie classic *It's a Wonderful Life*, and a fitting title it is. In the book, Gould describes the amazing diversity of life that appeared in what has come to be known as the Cambrian Explosion, and follows the chaotic way in which it evolved. Suddenly the fossil record in sedimentary rocks, very sparse up to this point, blossoms with an abundance of preserved creatures. Some of these are so bizarre that they challenge the imagination. How did they move? What did they eat? What did they actually do with those incredible appendages? A few of these wonderful beings are shown in Figure 7.3. But in spite of the great Cambrian expansion, life on earth had begun long before, probably more than two billion years before. It is to these very hazy beginnings, sometime in the early Archean, that we turn first.

IN THE BEGINNING

Philosophers and thinkers have expounded on the question of how life began for millennia. Some thought life to be eternal, without a beginning. Aristotle, who influenced thinking for two thousand years, believed that some life, and perhaps all, arose spontaneously. This idea, which was not unique to him, was based on observation: Plants suddenly appear in fertile soil after rains, and maggots materialize in decaying meat. In the 1920s, Aleksandr Oparin, a Russian biochemist, proposed and developed the idea that life originated in the warm, watery environment of the early earth's surface, under an atmosphere mostly composed of methane—the natural gas that heats our homes and cooks our food. The early seas were believed by Oparin to be rich in simple organic molecules, which reacted to form more complex molecules, eventually leading to proteins and life. Almost thirty years after Oparin published these ideas, Stanley Miller, who was then a graduate student at the University of Chicago with Nobel Prize winner Harold Urey, demonstrated that amino acids, the building blocks of the proteins necessary for life, could form under conditions thought to prevail on the early earth. Miller's experiment was elegant. He passed electric discharges through a mixture of methane, hydrogen, ammonia, and steam, and when he analyzed the results, found that he had made amino acids. The discharges were a proxy for lightning, the gas mixture an educated guess about what the early atmosphere may have been like. Amino acids cannot replicate themselves, and are not themselves alive. Nevertheless, this experiment has long been recognized as a landmark for understanding a process that must have been one of the important steps in the evolution of life on earth, the natural synthesis of amino acids. However, as we will see below, it now seems likely that the Miller-Urey experiments may not be directly applicable to the events of the early Archean.

One of the problems hindering understanding of the origin of life is that environmental conditions on the early earth are not known with any certainty. It is only possible to make reasoned estimates. For example, for some fairly long period of time after formation, perhaps as much as several hundred million years, the surface must have been much hotter than it is today. Continued impacts of meteorites, large and small, would have added further heat energy, and in the earliest part of earth history the larger impacting bodies may have broken through the cooling crust to expose underlying molten material. Large quantities of volcanic gases would have been released into the atmosphere as lavas erupted onto the surface, producing a greenhouse effect far more severe than anything likely to result from human activity. It is quite possible that the early atmosphere was many times as dense as today's, and that the seas and oceans were hot. Some have even suggested that because of the high atmospheric pressure, the oceans could have been hotter than the boiling point of water today—truly a pressure-cooker early earth. However, life as we know it is quite sensitive to temperature, and no modern organisms are known to survive much above 100°C. It is unlikely that life became established until surface temperatures had decreased to this level, or lower.

Although we don't know the precise composition of the early atmosphere, there has been enough progress made on this subject in recent years that it is possible to say with some certainty that the methane-rich composition envisioned by Oparin, and the methane-ammonia-hydrogen mixture used by Miller in his experiments, are probably not very realistic. Based on studies of our closest neighbor planets, Mars and Venus, and also considering evidence from the earth's sedimentary rocks, it seems probable that the earth's early atmosphere was rich in carbon dioxide rather than methane. On both Mars and Venus, CO_2 is by far the most abundant gas in the atmosphere. On the earth it is a minor constituent. But there is an enormous amount of this compound buried in the sedimentary rocks of the earth's crust, enough so that, if it were all released, our atmosphere would be much more like those of our neighboring planets. How did CO_2 gas end up in the crust? The answer lies in what geologists refer to as the carbon cycle. Through a series of chemical reactions, carbon dioxide from the atmosphere finds itself, in dissolved form, in the oceans. In seawater it combines with calcium to precipitate as calcium carbonate, the main constituent of limestone and the same material that clogs up water pipes and makes scales in teakettles. Over geologic time so much CO_2 from the atmosphere has been converted to limestone in this fashion that there is more than 100,000 times as much stored as limestone as there is in the atmosphere. A significant amount of carbon dioxide has also been extracted from the atmosphere by plants during photosynthesis, converted into organic material, and eventually buried and transformed into coal, oil, and natural gas. Burning these fossil fuels returns this carbon dioxide to the atmosphere, and is partly responsible for the much-discussed greenhouse effect and global warming.

In a carbon dioxide-rich atmosphere, the Miller-Urey electrical discharge method for creating amino acids doesn't work. If the early atmosphere was truly CO_2-rich, production of the necessary organic compounds for life must have occurred differently. Because we have no geologic record of the earliest events on our planet, the details of the processes are not known, and probably never will be. However, many plausible ideas have been suggested. Presumably, as on earth today, there would have been a myriad of microenvironments with different conditions of temperature, chemical composition, and energy supply. Furthermore, many organic compounds, even amino acids, have been found in meteorites. Radioastronomers have also identified organic compounds in interstellar space, and studies of Halley's comet during its most recent near-approach to the earth showed that it is rich in organic material. Inevitably, many organic compounds must have arrived on earth from space in the early Archean along with all of the other impacting material, and would have been dispersed over the surface. But life does not arise full-blown from such simple molecules, and there is still a giant step from these compounds to the production of complex, polymerized macromolecules and chemical systems capable of reproducing themselves. Various pathways that may have led to the beginnings of life are being actively investigated by chemists; for example, one line of inquiry indicates that surface chemistry may have been important. It is possible that the surfaces of naturally occurring materials, such as mineral grains, acted as templates for organizing and perhaps even replicating complicated molecules. However, in the absence of a detailed record, all that we can surmise is that over some extended period of time, reactions among a variety of successively more complex organic molecules eventually produced compounds and structures capable of reproducing themselves—at which point, life had begun.

At some very early stage in this process, membranes developed that permitted some of these organic molecules to segregate and accumulate in environments slightly different from those outside the membrane wall—in short, primitive cells were formed. In fact, the oldest fossils, the very earliest concrete evidence we have for life, are tiny preserved cells that resemble some modern bacteria. These objects occur in Archean sediments with an age of about 3.5 billion years.

 ## WHY DID IT TAKE SO LONG?

Although 3.5 billion years is very long ago by any standard, it is worth remembering that it is also a billion years after the earth formed. More than a fifth of the earth's history had already passed by. One of the reasons why there are no recognizable fossils older than 3.5 billion years may simply be that there are very few rocks older than this, and none at all older than about 3.9 billion years. In addition, all of the early Archean rocks that do exist have been through episodes of metamorphism that may have destroyed any evidence of life that they once contained. There are, however, hints that living organisms were present considerably before 3.5 billion years ago. The clues are contained in the ancient, 3.8-billion-year-old sediments from western Greenland that were mentioned in the previous chapter. Over their long lifetime, these sediments have been buried deeply, experienced strong heating and metamorphism, and eventually uplifted and exhumed so that today they are once again at the earth's surface. Their original features have been largely obliterated, and they contain no obvious fossils. However, they do contain accumulations of graphite—pure carbon, the stuff of life and of pencil lead. It is possible that the carbon originated inorganically, but it is more likely that it is a chemical fossil, a relic of organic compounds formed by organisms. The Greenland occurrence is not unique; graphite is also found in Archean rocks in many other parts of the world.

However, even 3.8 billion years ago is more than 700 million years after the earth formed. Over the most recent 700-million-year span of the earth's history, virtually the entire course of evolution, from single-celled organisms to whales, kangaroos, and man, has taken place. Many scientists believe that all of the necessary steps for the origin of life—formation of simple organic molecules from the constituents of the early oceans and atmosphere, polymerization of these molecules and reactions among them leading to more complex forms, and eventually the initiation of replication—could occur in a relatively short period of time, perhaps 10 million years or less. If this is so, why do we not have an earlier record of life? Did it really take the better part of a billion years to originate?

We have already noted that the earth's oldest rocks have invariably been so heated and folded during their long history that most traces of their original state have been obliterated, and that even if life did arise shortly after the earth formed, it may be that we simply don't have a preserved record. But there is also a reason to believe that life may have had a slow start. This has to do with the fact that the young earth was being bombarded by material from space.

Although tens of thousands of small meteorites fall on the continents each year, occasionally a much bigger body strikes. The mile-across Meteorite Crater in Arizona, a spectacular sight on a clear day for airline passengers coming and going from southern California, was formed by a moderate-sized meteorite that crashed into the earth about 50,000 years ago. In 1908, an impacting body that may have been a small comet exploded above a remote part of Siberia, flattening forests and creating a shock wave that was detected on seismograms thousands of kilometers away in western Europe. Intuitively, it is easy to realize that the influx of such natural space debris must have been much greater during the early days of the earth. After all, our planet was formed by agglomeration of material in orbit around the sun, and even when it had reached approximately its present size there would still have been a great deal of potential impact material left in its vicinity. In spite of this fairly obvious line of reasoning, it was not until the hard data from the Apollo missions to the moon became widely available that geologists generally began to appreciate the very important role that impacts must have played. The spectacular images of the fragments of comet Shoemaker-Levy 9 smashing into Jupiter in the summer of 1994, generating disturbances over areas of that planet as large as the entire earth, have only served to underscore this importance.

The moon, as can be seen even with a good pair of high-powered binoculars, has a pitted surface. At one time it was thought that many of these features were volcanic, but it is now known that nearly all are the result of impacts. The craters range in size from the great circular basins that form the dark-colored mare (named, incorrectly as it turned out, using the Latin word for seas), which may be 1,000 kilometers or more in diameter, to microscopic pits on rocks brought back by the astronauts. One of the many important results of lunar studies was the determination of the rate at which these craters formed. To nobody's surprise, the cratering rate on the moon was much higher during its early history than it is today. The largest craters, those now occupied by the mare, are the oldest. The graph in Figure 3.1 shows just how dramatically the rate of impact has slowed over the moon's lifetime.

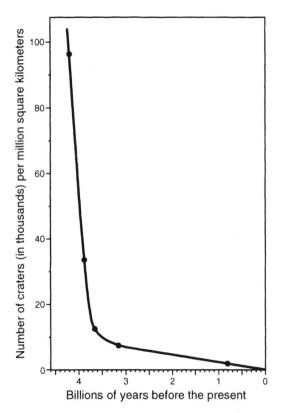

FIGURE 3.1 *The density of craters on various parts of the moon's surface has been measured using photographs taken from orbiting spacecraft. Some of these regions were visited during the Apollo missions, and samples were brought back to the earth and dated. This graph was constructed from such information, and it shows that the early moon—and by implication the nearby earth as well—underwent very heavy bombardment. Dots along the curve are the actual measured data points.*

The moon is a small planet that cooled quickly and has been geologically quiet for billions of years. There are no volcanoes and no earthquakes; there is also no atmosphere to cause weathering and erosion. In fact there are none of the geologic processes that constantly work to erase the rock record on the earth, and as a result the moon preserves a great deal of evidence about its early history. That evidence indicates that the moon formed at approximately the same time as the earth, and that virtually all lunar rocks greater than about 3.9 billion years of age, and many younger ones as well, have been severely altered by violent impact processes. The oldest parts of the moon's surface are completely saturated with impact craters. When the curve in Figure 3.1 is extrapolated to the early history of the moon, it shows that our closest neighbor was undergoing a relentless bombardment from space. If the moon was being pummeled by impacting bodies, the nearby earth, with many times the gravitational attraction of the moon, must have been even more severely battered.

What consequences would this have had for early life, just emerging on earth? The probable effects of the largest projectiles that may have struck our planet seem to be in the realm of science fiction, but in fact they may have occurred repeatedly over the first several hundred million years of earth history. An impacting body with a diameter of around 400 kilometers, about the size of some of the largest asteroids now in the asteroid belt, would vaporize itself and a substantial amount of the earth's surface, ejecting a gigantic plume of both vaporized and molten rock into the atmosphere. Some of this debris would actually be launched into space, but most would spread around the globe, heating both the atmosphere and the surface rocks to very high temperatures. It is likely that the entire existing ocean would have evaporated under the intense heat. The huge amount of water vapor injected into the atmosphere would have radically slowed the cooling process after the impact, because water is even more efficient for producing a greenhouse effect than is carbon dioxide. The surface of the earth would have been sterilized, and any primitive life in existence before the impact would almost certainly have been wiped out.

Even much smaller impacts would have had drastic consequences. The Imbrium Basin, the largest obvious impact feature that can be seen on the moon's surface, was formed by a body estimated to be about 100 kilometers in diameter. Imbrium is partly ringed by mountains about five kilometers high; the Apollo 15 landing site was at the foot of the Apennine Mountains, part of this ring. Samples brought back by the Apollo astronauts, as well as other evidence, show that the Apennine Mountains are unlike any mountain ranges on earth. They are simply a pile of rubble, part of the debris thrown out at the time of the Imbrium impact. An equivalent impact on earth would vaporize rock, create a huge crater, set off gigantic sea waves, heat the atmosphere, and probably lead to evaporation of at least the upper parts of the oceans. Life on land, or in the surface layers of the oceans, would be destroyed.

In addition to the body that created the Imbrium Basin, at least one other object with a diameter in the 100-kilometer range is known to have struck the moon during the first 600 to 700 million years of its existence. Thus there is a high probability that more and larger bodies struck the earth during this same period, raising the interesting possibility that the process of creation of living organisms from simple organic molecules may have occurred more than once on the earth, only to be reversed by the sterilizing effects of giant impacts. Chemists and biologists find it difficult enough to reconstruct the steps in the origin of life in a tranquil environment. Considering the periodically very violent episodes that almost certainly characterized the early earth, it is not surprising that the process would have been slow and sporadic.

Large, frequent collisions could also provide the answer to another puzzle. In the previous chapter, it was noted that if the sun evolved in the way that most stars of its size do, it would have been too weak during its early life to keep the earth from freezing over. Once frozen, calculations show, it would be difficult to thaw the earth even if the sun did warm up. However, the large impacts just described would provide periodic defrosting, preventing a permanent deep freeze before the sun's energy output increased toward its present-day value.

THE OLDEST FOSSILS

The oldest fossils date from 3.5 billion years ago. They were found in sedimentary rocks from northwestern Australia, and are microscopic, single-celled, bacteria-like organisms that appear to be very similar to a present-day group known as cyanobacteria. The fossils are in the form of filaments made up of many joined-together cells, as shown in Figure 3.2. The rocks in which they occur are finely layered sediments composed primarily of chert (fine-grained silica, or SiO_2) and are interpreted to have been deposited in a shallow-water environment, possibly a lagoon. Although simple, the fossils exhibit considerable variety in morphology, suggesting that such organisms may have arisen on the earth long before these particular sediments were laid down.

Bacteria were the unquestioned lords of the Archean. In fact, from their first appearance to the end of the Archean eon a billion years later, no other kinds of fossils are found. As we know well, bacteria are with us still, occupying every imaginable niche on the present earth. They are ours in sickness and in health, fostering infections and fermenting wine. It is difficult to imagine a world without bacteria.

Bacteria are single-celled organisms, but their cells don't contain a nucleus or some of the other internal structures characteristic of later, more advanced forms of life. In the modern world, some bacteria use energy from the sun to conduct photosynthesis, and in the process produce oxygen. Other bacteria use quite different kinds of

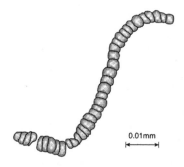

0.01mm

FIGURE 3.2   *A sketch of one of the oldest fossils ever found: a 3.5-billion-year-old bacterial filament from sedimentary rocks in north-western Australia. The sketch is based on photographs taken through a microscope. Redrawn from Figure 1.5.5 (A₂) by J. W. Schopf, in* **The Proterozoic Biosphere,** *page 31, edited by J. W. Schopf and C. Klein. Cambridge University Press, 1992. Used with permission.*

chemical reactions to grow and reproduce. Exactly when in the history of life photosynthesis developed is a controversial but important question, because it bears directly on the evolution of the atmosphere from an early one dominated by carbon dioxide to something closer to today's breathable, oxygen-rich one.

In rocks that are not quite 100 million years younger than those containing the first microscopic fossils, much larger remains of living organisms appear, fossils that can be seen easily with the naked eye. These are peculiar bulbous structures that look a bit like very large, layered cabbages, and can reach several meters in height. But appearances are deceiving. These objects, called stromatolites, are not single organisms at all, but are essentially colonies of bacteria. They are made up of individual cells of cyanobacteria similar to those of the earliest fossils.

Fossil stromatolites become increasingly more abundant in younger sedimentary rocks, and by the end of the Archean and into the following Proterozoic eon they are fairly common and quite conspicuous. Their peculiar shape results from the fact that they are formed by layer upon layer of bacterial mats, which trap sand and assorted particulate material in their sticky, filamentous strands. In spite of the fact that they are among the earliest fossils known, stromatolites are still found today as colonies of living organisms, although they are not nearly as widespread as they were in the Proterozoic. They grow in shallow water in tropical environments, suggesting that the ones we find fossilized in Archean rocks grew in the coastal regions of Archean continents.

The colonies of cyanobacteria that make stromatolites today live by photosynthesis. Although this does not prove that their Archean counterparts were also photosynthetic, it does indicate that by about 3.5 billion years ago, photosynthesis could already have been established on the earth. Nevertheless, there is little evidence that the atmosphere contained much oxygen even by the end of the Archean eon. However, as we will see in the next chapter, that began to change early in the Proterozoic eon.

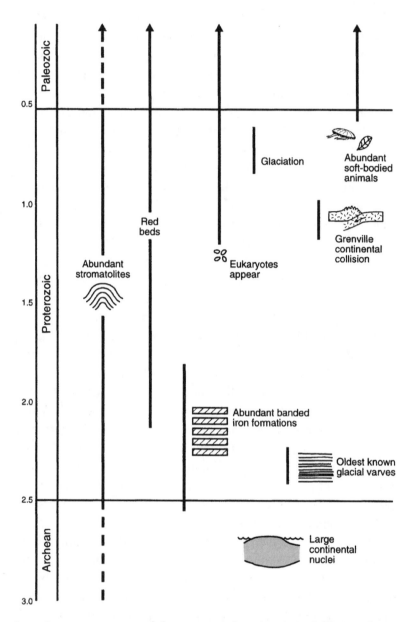

Some important events of the Proterozoic. Time is in billions of years before present.

4

THE
PROTEROZOIC EON

LIKE THE ARCHEAN, the Proterozoic eon lasted almost two billion years. By its end, nearly nine-tenths of the earth's 4.5 billion years of history had gone by. Although we know considerably more about the Proterozoic than we do about the Archean, the record is still very incomplete, particularly for the early parts. However, Proterozoic rocks are relatively abundant compared to their Archean counterparts. We know from them that stromatolites became quite common, that the oxygen content of the atmosphere increased, and that mountain ranges much like today's were being formed and then destroyed. We even know a bit about the climate. What are the clues that provide such information? Perhaps this is a good place to examine some of the ways in which geologists read the records in the rocks, using examples from the Proterozoic.

A fundamental concept in the earth sciences is the principle of uniformitarianism. The word means what it says. In textbooks, the phrase that is often used is "the present is the key to the past." Actually, there is nothing unique to the earth sciences about this concept; it just emphasizes the fact that geologic processes are governed by the same laws of physics and chemistry, and have the same kinds of mathematical descriptions, as everything else in nature. If a 300-million-year-old sandstone exhibits ripples that are similar to those forming in sand on a protected beach today, then chances are it was deposited in just such an environment. Although the concept of uniformitarianism may seem obvious, the idea was a revolutionary one in its time. It was the Scottish geologist Hutton who first applied it systematically to studies in the earth sciences. The concept has had its detractors, but when used with common sense, and a view to the very long spans of time involved, it serves geology well. Even events that are rare and catastrophic from a human point of view, such as a once-in-a-century flood, or a devastating earthquake, or even a large meteorite impact, are actually regular, periodic, and to some extent predictable events on geologic timescales.

We've learned that the earliest atmosphere was rich in CO_2, and that even at the end of the Archean there was probably very little oxygen in the atmosphere. But rocks from the Proterozoic tell a different story, and a fascinating tale it is. By examining these rocks in detail, and at the same time taking into account the ideas of uniformitarianism, geologists can reconstruct at least some of the steps toward the present-day atmosphere.

EVOLUTION OF THE ATMOSPHERE

The clues that Proterozoic rocks contain about the changing composition of the atmosphere suggest that there was a dramatic increase in oxygen concentration during the eon. We know that the present-day content, which is sustained by photosynthesis in plants, is about 21 percent by volume, and it is clear that past variations in atmospheric oxygen levels have been inextricably tied to the history of life on the earth. There are some interesting and unexpected consequences of changing atmospheric oxygen that we will examine below—such as where we get iron ore for our steel foundries.

A peculiarity of some early Proterozoic sedimentary rocks with ages of more than about two billion years is that they contain the minerals pyrite (also known as fool's gold), and uraninite. The chemical composition of pyrite is iron sulfide, FeS_2, and, as you might guess, uraninite is a uranium mineral. In a few places in the world the uraninite concentration of Proterozoic rocks is high enough that it is mined as uranium ore. In itself, the occurrence of these two minerals is not particularly remarkable—they are also found in rocks of other ages. What sets the early Proterozoic pyrite and uraninite apart, however, is the fact that they occur in sediments that were originally deposited in environments such as streambeds and beaches. Careful examination shows that the minerals themselves are detrital grains, eroded out of some parent rock and carried to their place of deposition by flowing water. However, neither uraninite nor pyrite are found in such environments today, because they are unstable in the presence of oxygen. In a very short time they oxidize and break down. Presumably, the rivers or streams in which these grains were transported in the Proterozoic were in contact with the atmosphere, as are streams today. Uniformitarianism thus suggests that something was very different in the early Proterozoic; the obvious answer is that the atmosphere contained so little oxygen that both uraninite and pyrite could survive as detrital grains without being oxidized. These minerals are not found in stream sediments with ages less than about two billion years, indicating that atmospheric oxygen began to increase at about that time.

It is always possible, although perhaps unlikely, that the uraninite and pyrite grains were preserved from oxidation by some yet-to-be-discovered mechanism. But there are at least two other clues in Proterozoic rocks that also suggest that the atmosphere had a low oxygen content until about two billion years ago. One of these is related to the mining of iron.

Much of the world's iron ore comes from deposits known as banded iron formations, BIFs, for short. The ore is found in sedimentary rocks, and the deposits are banded, typically with alternating layers of iron-rich and silica-rich rock. The iron-rich layers are much darker than the silica-rich layers, giving the deposits their striking banded appearance. Most of the world's BIFs occur in the early part of the Proterozoic, with ages greater than about 1.8 billion years.

Understanding the significance of the BIFs as indicators of atmospheric oxygen contents requires a bit of knowledge about the chemical behavior of iron, which is quite dependent on the amount of oxygen in its surroundings. Iron metal, as anyone who owns a car knows, combines with oxygen very rapidly, forming rust. But iron in ordinary rocks of the earth's crust is not present as metal. Rather it exists in one of two valence states (also called oxidation states), either Fe^{2+} or Fe^{3+}, and is combined with other elements to form the typical minerals found in common rocks. In igneous rocks, most of which originate by melting of the mantle, much of the iron is in the lower oxidation state, Fe^{2+}. However, when these rocks are weathered by rainwater some of this iron dissolves, and the high oxygen content of the atmosphere very quickly causes it to be oxidized to Fe^{3+}. But Fe^{3+} is quite insoluble, and the iron thus precipitates very rapidly as a fine-grained, rust-like substance that leaves reddish stains on stream bottoms or anywhere else it collects. As a result, natural waters on the earth today contain very little iron in dissolved form. On the other hand, if the oxygen content of the atmosphere were much lower, the Fe^{2+} would not be oxidized and these same waters could hold a great deal more dissolved iron, because Fe^{2+} is much more soluble than Fe^{3+}.

The BIF deposits were laid down in water, and the geology of most of them suggests that they formed in seas quite close to land, although in a variety of water depths. The iron in these sediments is oxidized iron, Fe^{3+}, precipitated from the overlying waters. Because there is evidence that atmospheric oxygen was still low when the deposits were formed, it has been inferred that the oxygen necessary for this process was produced by photosynthetic algae living in the surface waters of the seas. But the important question that bears on the composition of the atmosphere is: How were the huge amounts of iron in these deposits transported to the site of deposition in the first place? As implied above, under present-day conditions very little of the iron dissolved from rocks on land is carried to the oceans. Instead, it is rapidly oxidized and precipitated out as iron oxides. The same is true for iron dissolved from the basaltic rocks of the seafloor by the circulating waters of undersea hot springs. This once again suggests that conditions were different in the early Proterozoic. Low atmospheric oxygen contents would have permitted transport of quite large amounts of iron as Fe^{2+}. When it encountered zones of surface seawater that were relatively oxygenated by photosynthetic algae, it would have precipitated from solution as iron oxide. The fact that most known banded iron formations are restricted to that part of the geologic timescale before about 1.8

billion years ago suggests that by then the oxygen content of the atmosphere had increased to the point where large amounts of dissolved iron could no longer be transported.

The third piece of evidence bearing on atmospheric oxygen also involves the oxidation of iron. In the geologic record, accumulations of sediments with distinctive reddish coloration, usually sandstones, occur quite commonly. Not surprisingly, geologists refer to these rocks as red beds. The color comes from the presence of fine-grained oxidized iron, in the form of the mineral hematite, which commonly coats and sometimes cements together the individual grains of the sandstone. Red beds are often quarried for building stone, as anyone who has seen the Red Fort in Old Delhi, or the cathedrals at Chester or Carlisle in the northwest of England, can attest. No red beds older than about 2.2 or 2.3 billion years are known, probably because there was insufficient oxygen in the atmosphere before then to produce hematite cements. Once again it is necessary to add the caveat that there may be other reasons for this lack. For example, some geologists have pointed out that the types of environments in which red beds are deposited may not have been common in the Archean or the early Proterozoic. Many red beds are composed of nonmarine sediments, laid down within large continents under arid conditions, and the small continents typical of the earliest part of the geologic record may not have been conducive to such deposits. However, there *are* sediments with ages greater than 2 billion years that appear to have been formed under conditions that today would result in red beds, yet they are not cemented with hematite. The weight of the evidence points to atmospheric oxygen as the determining factor in their occurrence.

Thus even the imperfect geologic record of the Proterozoic gives us some very powerful knowledge about the way in which the earth's atmosphere has evolved. It suggests that around two billion years ago, give or take a few hundred million years, there was a marked increase in the oxygen content. After this time, uraninite and pyrite grains could not accumulate as detrital grains in rivers and beach sands; they were oxidized and broken down. Iron dissolved from both continental and seafloor rocks was quickly oxidized and precipitated, and the massive amounts needed to form banded iron formations could no longer be transported to, or even within, the sea. And for the same reason, hematite could precipitate from the intergranular water in sandstones, forming coatings and cement, and creating extensive red beds during the rest of the geologic record. Although the individual clues are never unequivocal, the cumulative evidence is very persuasive. Like detectives, geologists have amassed bits of seemingly disparate information that together show beyond reasonable doubt the details of events that occurred more than *two billion years ago!*

In spite of the fact that the fossil record in Proterozoic rocks is quite scanty, it corroborates the conclusions reached about atmospheric oxygen from other evidence. The record indicates that it was not until late in the Proterozoic that complex, multicelled organisms developed, but it also shows that the stromatolites became very prominent early in the eon. Modern stromatolites live in tropical intertidal zones and are essentially colonies of oxygen-producing photosynthetic algae. It is possible that shallow coastal or inland seas on the relatively large continents that developed in the late Archean and early Proterozoic provided just the right environment for stromatolites to flourish, leading to much higher rates of oxygen production than had occurred earlier. Because it is such a reactive substance, however, most of the free oxygen initially produced by photosynthesis would have been used up quickly in chemical reactions as both the components of surface rocks and various constituents of the atmosphere itself were oxidized. Eventually, though, as the rate of photosynthesis increased, oxygen began to accumulate in the atmosphere.

 ## THE PROTEROZOIC CLIMATE

There are only a few clues in the rock record about climate in the Proterozoic. Much of our information about climate in the more recent parts of geologic history comes from the fossil record, because we have a reasonably good understanding of the types of environment in which many fossil organisms flourished. The scarce fossils of the Proterozoic, mostly single-celled bacteria, provide little evidence in this regard. However, the rocks do include the earliest evidence for glaciation, probably a global ice age.

The inference that some types of sedimentary rocks are the result of glacial activity is based on the principle of uniformitarianism—the deposits associated with present-day glaciers have been well studied, and some of their characteristics are quite distinctive. In 2.3-billion-year-old rocks in Canada near Lake Huron, there are thin laminae of fine-grained sediments that resemble varves, the annual layers of sediment deposited in glacial lakes. Typically, present-day varves show a two-layered annual cycle, one layer corresponding to the rapid ice melting and sediment transport of the summer season, and the other a finer-grained layer corresponding to slower winter deposition. Although it is not easy to discern such details in the Proterozoic examples, they are almost certainly glacial varves. These fine-grained, layered sediments even contain occasional large pebbles or "dropstones," a characteristic feature of glacial environments where coarse material is sometimes carried on floating ice and dropped far from its source, into otherwise very fine-grained sediment. Glacial sediments of about the same age as those in Canada have been found in other parts of North America, and in Africa, India, and Europe. This indicates that the glaciation was global, and that for a period of time in the early Proterozoic (the duration is not well known) the earth was gripped in an ice age.

Although there are many areas of the earth's crust with rocks older than 2.3 billion years, none of them shows clear evidence for earlier periods of glaciation. This does not mean they didn't occur, for the record is full of gaps, and most of the older rocks are highly metamorphosed so that their histories are difficult to decipher. Nevertheless, the evidence is suggestive that the 2.3-billion-year glaciation was one of the first major periods of deep freeze the earth experienced, at least after the rock record began to accumulate starting around 3.9 billion years ago. (The possible freeze-over of the oceans, mentioned in Chapter 2, is an event quite different in scale and cause from the glaciations discussed here, and in any event if it occurred at all, took place long before the rock record begins.) Following the early Proterozoic glaciation, however, the climate appears to have been fairly benign for a very long time. There is no evidence for glaciation for the next 1.5 billion years or so. Then, suddenly, the rock record indicates a series of glacial episodes between about 850 and 600 million years ago, near the end of the Proterozoic eon. These later periods were also global phenomena, for all of the present continents (with the possible exception of

Antarctica, much of which is buried in ice now and not accessible for investigation) show evidence for glaciation during this time. Although the continents were arranged quite differently in the late Proterozoic, the widespread distribution of evidence for glaciation suggests that much of the planet may have been cold, even at low latitudes. Winter vacations in the late Proterozoic equivalent of the Caribbean would not have been much fun.

 ## THE EVOLUTION OF CONTINENTS

And what were the continents like in the Proterozoic? Earlier, it was pointed out that in the early Archean they were mostly small and probably not very much like their present-day counterparts. Toward the end of the Archean, larger continents were in existence, and by the end of the Proterozoic their size and physical nature were much more like that of today's continents. There is ample evidence of continent-building events during the long span of Proterozoic history, and it suggests that the processes were not too dissimilar from those of the present. One of the best-documented examples is an area of northern Canada that has been studied by Paul Hoffman, of the Geological Survey of Canada.

Hoffman spent summers mapping the rocks that outcrop in the Northwest Territories. Over a vast region straddling the Arctic Circle and stretching from the northern shores of mainland Canada to Great Slave Lake in the south, he recognized the remains of a Proterozoic cycle of erosion, sedimentation, and mountain building (see Figure 4.2). The Proterozoic mountains have long since eroded away, and indeed the landscape today is subdued and bleak. But it has a haunting beauty of its own, and, best of all for a geologist, much of it has very little vegetation. The rocks are bare and exposed, ready to tell their tale.

But how is it possible to put together from mere rocks a story about what happened more than two billion years ago? We have already had a glimpse of the process in the discussion of atmospheric oxygen, and to do the subject justice would take an entire book in itself. Interpreting the evidence requires a deep understanding of geology, together with much experience in the field pondering over rocks. However, some of the basic elements are quite simple, and are really just applications of common sense. Take the example of time. More will be said of this in a later chapter, but it is quite obvious that time, especially in the form of the ages of rocks and the rates of various geologic processes, is critical for understanding the geologic history of a region. And at least relative time, the question of whether a rock or a rock formation is older or younger than its neighbors, can often be worked out very simply. For example, in sequences of sediments, the oldest deposits are generally at the bottom of the pile, the youngest at the top. For other rocks, crosscutting relationships are often the key: If an igneous body, or a fault, cuts across another rock formation, the crosscutting feature is clearly the younger of the two. Such examples may seem simplistic, but by applying just this kind of reasoning it is often possible to work out relative age relationships even in very complex situations (see Figure 4.1). Only when that task is done is it possible to reconstruct the actual geologic events in their correct sequence.

To return to the Proterozoic rocks in northern Canada, Hoffman found that the region he was investigating had been the edge of a continent in the early Proterozoic, shedding sediments rich in quartz toward the sea (see Figure 4.2). Quartz sand is a good tracer of continents; weather a granite, the typical rock of the continental crust, and you get lots of quartz. Most of the other minerals in the granite dissolve, or are transformed to something else, for example, clay. The white sands of tropical islands (most of which are coral-fringed volcanoes with very different compositions from the continents) may look like the beaches of California or Spain, but they are made of bits of coral, not quartz. The quartz-rich Proterozoic sandstones mapped by Hoffman show that the continental sediments came from the east, and that the ocean lay to the west—at least in terms of the present-day geography. The actual orientation of the continent may have been quite different in the Proterozoic. But higher up in the sediment column, and therefore at a later time, different sediments appear, with volcanic material in them. And in contrast to the quartz sands, the volcanic sediments came from the west, seaward of the continent. How can this be? Before the theory of plate tectonics came into being, puzzles such as this were solved by suggesting that there must have been a "missing" landmass somewhere out to sea. We now recognize that indeed there was land seaward of the continental margin, but we

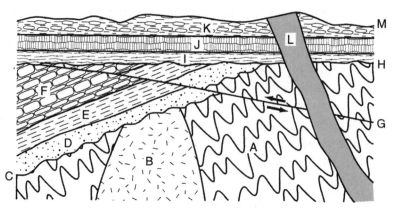

FIGURE 4.1 *Geologic cross sections can contain a tremendous amount of information, although working out time relationships among the various rock units can be like solving a puzzle. Can you solve this one? The actual sequence of events is indicated by the lettering, as follows: A, deposition of sediments, then metamorphism and folding; B, intrusion of a granitic magma into the metamorphosed sediments; C, an erosional surface developed on units A and B by weathering at the earth's surface (this indicates that A and B must have been uplifted, because metamorphism of A and intrusion of B both occurred deep in the crust); D–F, deposition of sedimentary layers from a body of water; G, development of a fault (note that the fault does not cut through units younger than F, and is now inactive); H, another erosional surface (note that since units D, E, and F, like all sediments, were horizontal when they were deposited, the whole area must have been tilted before the erosion occurred. There could be a very large time gap between F and I); I–K, further deposition of sedimentary units; L, intrusion of an igneous rock body, probably one that fed lava flows on the surface, which have since been eroded away; M, the present-day surface, produced by erosion.*

can infer from knowledge of how things operate today that it was probably a group of volcanoes much like the Aleutians or the Mariana Islands that was being carried toward the continent and eventually collided with it. There is no equivalent of the Proterozoic ocean in the Northwest Territories today; the western edge of the continent is more than 1,000 kilometers away.

This example is by no means unique. Collisions between landmasses over geologic time, suturing them together along moun-

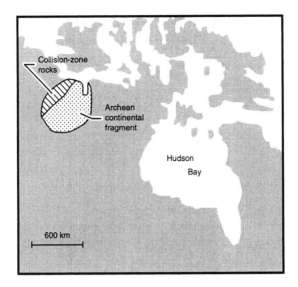

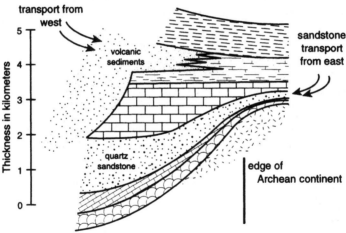

FIGURE 4.2 *Paul Hoffman of the Geological Survey of Canada mapped the Proterozoic rock units found along the western edge of an Archean continental fragment in northern Canada (upper diagram). Although the sedimentary rocks are now faulted and metamorphosed, Hoffman was able to reconstruct the sedimentary sequence (lower diagram), which indicates that sediments eroded from the continent to the east accumulated along its margin, then, later, volcanic material began to appear from the west, indicating the approach (and eventual collision) of another continent and/or island arc. Modified after Figures 10-1 and 10-4 in* Earth and Life Through Time, *2nd edition, by S. M. Stanley. Copyright © 1989 by W. H. Freeman and Company.*

tain ranges—and sometimes also the reverse process, continental breakup—have led to the present configuration of land and sea. North America, one of the largest continents, is typical; in many ways it resembles a giant patchwork quilt, assembled from fragments of disparate material.

The picture painted above for the rocks of the Northwest Territories—first sandstone being deposited along the margin of a continent, its source to the east, then volcanic sediments from the west—is greatly simplified. In fact, these rocks have been metamorphosed, folded, and cut by numerous faults, making the job of reconstructing their original disposition an extremely difficult one. The folding, faulting, and metamorphism are almost certainly the result of collision of the continental and volcanic blocks, and the mountain-building episode that accompanied it. In all aspects—the type of faulting, the bands of metamorphic rocks that parallel the ancient coastline, and the types and sequences of rocks—this region resembles more modern zones of collision and mountain building. But as already pointed out, there are no mountains present in this region today, only low-lying, subdued topography. Again we are reminded that on geologic time scales the earth is a very dynamic place.

In mountainous areas, erosion carries away 1 to 1.5 meters of material every 1,000 years. At that rate, even Mt. Everest would be worn down to sea level in five to eight million years. However, things are not quite so simple, because as a mountain is eroded away and its slopes become less steep, the rate of erosion slows down also. Partly for this reason, Mt. Everest and the rest of the Himalayan mountains will be around (although in much subdued form) for considerably longer than their present-day erosion rates would suggest. But even more important is the fact that mountains are a bit like ships floating on the sea: Remove some of the cargo, and the ship rides higher in the water. Similarly, when material is removed from a mountaintop by erosion, the crust will "float" up a bit higher in the underlying mantle. If erosion removes a meter of rock, the earth's response to the reduced weight is uplift, and the actual decrease in elevation is only about 20 centimeters. For this reason, it probably takes 50 to 60 million years for a typical large mountain range to be planed down close to sea level, still not a particularly long time in geologic terms. The Rockies, the Alps, the Himalayas—all will eventually disappear, but they will leave behind a telltale record of their formation in the rocks that remain.

The event that produced the now-vanished mountain range in the Canadian Northwest Territories took place close to 1.9 billion years ago. But it was just one of many such collisions. By 1.6 billion years, about halfway through the Proterozoic eon, much of what is now North America had been assembled from smaller fragments into a supercontinent that geologists who have studied these rocks call Laurentia. Paul Hoffman wrote a paper about the process, and titled it "The United Plates of America." The giant continent of the middle Proterozoic also included Greenland and the northern parts of the British Isles, as well as bits of Scandinavia and northern Russia.

In other parts of the world, similar events were taking place. Most of today's continents contain small fragments of Archean crust, sutured together at collision zones to other Archean or Proterozoic fragments. And it is possible, although not yet proven, that virtually all of today's continents were united in a truly giant continent near the end of the Proterozoic. Part of the evidence comes from a belt of metamorphic rocks that runs down eastern North America from Labrador to the Gulf of Mexico. These rocks have ages between 1.2 and 1.0 billion years, and are collectively called the Grenville Province (see Figure 4.3). They are exposed at the surface in eastern Canada and in the Adirondacks of New York State, but are also present, although buried, throughout much of the eastern United States. The Grenville rocks are a remnant of a massive collision between two large continents, with what we now recognize as North America on the west, and quite possibly South America—itself joined with most of the other continents—on the east. This marriage of North American and another large continent lasted several hundred million years, until they again began to rift apart about 800 million (0.8 billion) years ago—still in the Proterozoic. This rifting left a band of Grenville rocks behind along the eastern margin of North America. As we will see in Chapter 8, yet another strip of continent was added to the eastern side of North America even later, in a process much the same as the one that created the Grenville Province. That strip we know as the Appalachians. These various fragments of crust that now comprise the North American continent are shown in map form in Figure 4.3.

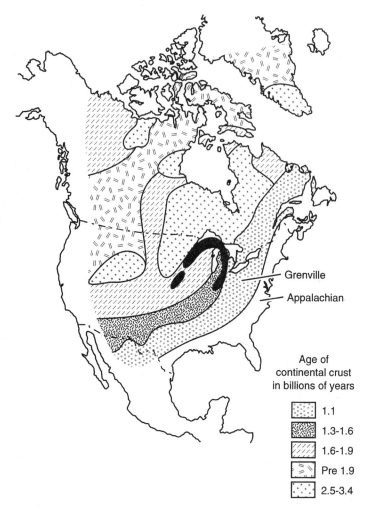

Grenville

Appalachian

Age of
continental crust
in billions of years

░	1.1
▒	1.3-1.6
▨	1.6-1.9
▨	Pre 1.9
░	2.5-3.4

FIGURE 4.3 *This generalized age map of North America, based on many hundreds of individual age determinations, shows that the continent is made up of several large crustal pieces, and in a general sense becomes younger outward. Both the Grenville and Appalachian Provinces record Himalayan-type mountain-building episodes, when large continents to the east collided with North America, only to split away again at a later time. Much of the material added during these collisions was in the form of sedimentary rocks, fragments of volcanic island arcs, or parts of the seafloor that originally separated the continents, although sometimes portions of the colliding landmasses were also left behind when they later separated. Figure 8.2, on page 122, shows how the process may have operated during the formation of the Appalachians. The dark gray horseshoe-shaped feature that extends through most of Lake Superior is the failed midcontinental Proterozoic rift discussed in the text.*

Actually, the assembling North American continent almost didn't survive the Proterozoic intact. Arched through the Lake Superior region, with two arms pointing southward into the midcontinent, is a great horseshoe-shaped scar in the continental crust (see Figure 4.3). It is a rift, an aborted continental sundering that occurred between 1.3 and 1.2 billion years ago. Although the rift is now filled in, it is recognized by the rock types that occur in it: basalts, which characteristically erupt in places where the crust is pulled apart, and sediments that are typical of those that fill in rift valleys. In some places, such as around Lake Superior, these rocks are exposed at the surface, and in others they are buried and have been recognized only in drill cores. Also, because the basaltic rocks of the rift have high densities and high iron contents, both the regional gravitational field and the magnetic pattern are strongly influenced. The location of the rift can thus be recognized from geophysical measurements made with instruments on the surface, even in places where it has been completely covered over by later sediments. What could have caused this huge rift, almost 2,000 kilometers long and more than 100 kilometers wide in places, and containing enormous volumes of basalt lavas? Almost certainly it was the result of a plume of hot material welling up through the mantle to impinge on the North American continent. Such features today, rising under the oceans, are responsible for the voluminous volcanism of Hawaii and Iceland. They are discussed in more detail in the next chapter. But North America proved too resilient to be broken up even by a mantle plume, and although scarred, it resisted fragmentation.

THE PROTEROZOIC BIOLOGICAL REALM

To the best of our knowledge, during much of the Proterozoic, while entire continents were forming, colliding, and rifting apart, remarkably little change was taking place in the biosphere, the realm of living things. The biological action, such as it was, took place largely in the oceans, and most of it occurred near the end of the Proterozoic. Even at the beginning of the Cambrian period the continents still weren't a haven for life. Although there may have been algae and perhaps even some primitive multicellular organisms living on the continents, the land surface was a very barren place compared to today.

We learned in the previous chapter that there are rare Archean fossils that are single-celled organisms. They appear to be bacteria and cyanobacteria (also known as blue-green algae), cells that don't contain nuclei or other important internal structures that characterize more advanced forms of life. They have been termed *prokaryotes.* Stromatolites are constructed by prokaryotes, and we have noted that they are perhaps the most characteristic fossils of the Proterozoic. Prokaryotes seem to have been the sole inhabitants of the Proterozoic seas until about halfway through the era. But then something remarkable occurred. The next step toward complexity, the development of *eukaryote* cells, with various internal structures, is now generally believed to have taken place when one prokaryotic cell engulfed another, intending, it is presumed, to consume it. Instead, the engulfed cell carried on, living in happy symbiosis and being modified along the way. A good example is the chloroplast, the structure in single eukaryote cells and higher plants where photosynthesis actually occurs. Chloroplasts resemble nothing so much as slightly modified cyanobacteria, or blue-green algae cells. Cells with internal structures, almost certainly eukaryotes, first appear in the fossil record about 1.4 billion years ago.

Surprisingly, even after eukaryotic cells developed, there was no immediate explosion of multicellular animals. That took many hundreds of millions of years—much longer than the time from the first appearance of dinosaurs on the earth to the present. A few fossils that appear to be multicellular algae occur in rocks as old as 1.3 billion years, but no traces of multicellular animals have been found in rocks older than one billion years. And even then, further development was quite slow until just before the "Cambrian explosion," which is described in a later chapter. Why did it take so long for complex life forms to develop on earth? That is a question that stumped Darwin, even though he didn't realize how truly vast was the span of time before the Cambrian. It continues to puzzle those who are concerned with the evolution of life. Certainly a part of the answer may be the imperfection of the fossil record before the Cambrian. Until then, organisms had not developed the hard, mineralized body parts—teeth, external carapaces, skeletons— that resist the effects of predators and are preserved in the rocks relatively efficiently. All examples of pre-Cambrian life are soft-bodied. Indeed, until the 1950s, paleontologists had not uncovered *any* undisputed evidence for life before the Cambrian, in spite of energetic efforts to do so. We may still be missing some of the crucial steps in the development of higher creatures. But even so, the early development of life was clearly a very slow process compared with the later pace of evolution. The reason why is not yet known, making this just one of the many mysteries that make the study of the earth's history so fascinating.

5

DANCE OF THE PLATES

THIRTY OR FORTY years ago some of the ideas expressed in the previous chapter, in particular the notion that continents rifted and sutured during the Proterozoic, would have seemed outrageous to most geologists. Today such descriptions are taken for granted. The development of plate tectonics in the intervening years has completely changed the way in which geologists view the earth. Before continuing our journey through geologic history, it is worthwhile to examine briefly how the theory of plate tectonics evolved, and to explore our current understanding of the movement of continents on the earth's surface.

Most people who have looked thoughtfully at a world map, centered as it often is on the Atlantic Ocean, have noticed that the coastlines of Africa and South America look as though they would fit neatly together if the Atlantic Ocean were removed. In spite of the fact that thousands of people must have made this observation, it was not until the early part of this century that its implications were seriously pursued. It was then that Alfred Wegener, a German meteorologist, began to gather and compare information about the flora and fauna of the continents. He also carefully examined what was known about their geology and paleontology, or fossil record. With the knowledge he acquired, Wegener came to the inescapable conclusion that various continents, including South America and

Africa, had been joined together at some time in the past. He discovered, for example, that some geologic features that appear to terminate abruptly at the coast of South America have counterparts in Africa, and when he fit the continents together like pieces of a jigsaw puzzle, the features were continuous. He also found that there is geologic evidence for ancient glaciation, at roughly the same time, in Australia, India and southern Africa. Once again he discovered that he could fit the continents together in such a way that the glaciated areas were continuous. In 1915 he published a book (in German) called *The Origin of Continents and Oceans,* in which he discussed this evidence in great detail and proposed his theory of "continental drift." However, in spite of the mass of geological data Wegener had accumulated, he glossed over many important details and was quite selective when he chose evidence to support his case. Partly for this reason his hypothesis was not taken seriously. Furthermore, prominent physicists of the day proclaimed that the outer part of the earth is much too rigid to permit the continents to drift about like ships on the sea. In particular, they pointed out that the forces Wegener called upon to move the continents—the centrifugal forces resulting from the earth's rotation—are much too weak to do the job. Wegener's ideas foundered for lack of a mechanism: Without a driving force, it was said, continental drift can't occur.

However, Wegener was on the right track. Although it doesn't occur precisely as he envisioned, continental drift is a reality. Just as Wegener proposed, Africa and South America were indeed joined in the past. At least once in the earth's history, *all* of the present continents were linked to form a supercontinent that stretched from pole to pole. Wegener's continental drift is discussed in textbooks and taught in high school, and it forms the underpinnings of much of what is understood about how the earth works. Today it is called plate tectonics.

EVIDENCE FROM THE OCEAN'S BOTTOM

The rebirth of Wegener's ideas as the theory of plate tectonics came about largely through studies of the seafloor carried out in the 1950s and 1960s. During and after the Second World War, the

United States Navy was anxious to know as much about the oceans as possible. Geologists and geophysicists obliged, some perhaps for patriotic reasons but many because they recognized the navy's interest as a golden opportunity to learn about the seafloor. At that time it was a scientific frontier and was virtually unknown; even much later many geologists were fond of saying that we know more about the face of the moon than we do about the ocean's bottom. The navy was generous, and oceanographic research expanded rapidly. Much of it was unclassified, and the discoveries that were made jolted the earth sciences into a new and more quantitative understanding of the earth.

One of the striking results of the intensive studies of the seafloor was an improved knowledge of its topography. Some information was already available, of course, gathered over the long history of sea travel. The earliest measurements were made very simply, by throwing a sounding line over the side of a ship and measuring the amount paid out, but such data were limited to shallow, near-shore regions that were heavily traveled. Echo sounders, which first made their appearance on ships in the 1920s but were not very sophisticated nor widely used until much later, were responsible for the wealth of information gathered in the 1950s and 1960s. These instruments measure, quite precisely, the length of time it takes a pulse of sound to travel from the ship to the seafloor and back again. Because the speed of sound in seawater is well known, it is a simple matter to calculate the depth. The beauty of echo sounders was that they could be kept going, day and night, no matter what the ship was doing. Every oceanographic expedition ran its echo sounder continuously, and the details of the seafloor began to emerge.

Today it's even easier to chart the ocean's bottom—it can be done by satellites without ever taking a ship to sea. The satellites measure the "height" of the sea surface very accurately. When variations due to tides and waves are factored out, a remarkable image emerges. The differences in the level of the sea surface from place to place actually map out the topography of the seafloor. The reason is that small variations in gravitational pull due to features on the seafloor—for example, the extra mass of a large volcano, or alternatively the mass deficit of a deep trench—actually affect the level of the sea surface above them. This relatively new technology has revealed some features that had never been discerned by measurements from ships.

But to return to the information about seafloor topography gathered by oceanographic ships in the 1950s and 1960s: It soon became quite clear that the ocean's bottom was not as monotonous as many had imagined it to be. Conventionally, the deep oceans had been thought of as geologically quiet, unchanging places where layer upon layer of the fine mud and silt washed off the continents had been accumulating since the beginning of time. Few people had thought this through very carefully, because the amount of sediment in the oceans would be enormous if this were really the case. However, as the data about the ocean floor came in, it quickly became apparent that instead of a flat, featureless deep-sea floor blanketed in sediments, there were huge ridges, deep trenches, great volcanoes, and long, steep escarpments on the seafloor. The immediate challenge was to understand how such features were formed.

Many people have seen the popular world maps, first produced by the National Geographic Society, that show the earth's surface with the oceans drained dry. Although they are somewhat idealized, the most striking feature about these maps is the extensive ridges, or rises, that appear on the seafloor. It has been said that if the oceans were removed, these features would be the most obvious characteristic of the earth's topography that could be seen from space. The ridge in the Atlantic is especially prominent on maps, again at least partly because the Atlantic Ocean is usually at their center. The Mid-Atlantic Ridge quite accurately bisects the ocean, following the bulges and indentations of the shorelines on either side, and it thus also roughly bisects the map. On average, it stands about 2.5 kilometers above the deeper parts of the seafloor to the east and west, and in most places it has a rift down its center. In the North Atlantic, the ridge rises above the sea surface as the island of Iceland.

The ridge that bisects the Atlantic is actually part of a more-or-less continuous system that extends into all oceans. It circles the Antarctic continent, and extends in a couple of branches into the Indian Ocean and up into the Arabian Sea. It wanders up the eastern part of the Pacific Ocean, and seems to come to a dead end near Baja California in Mexico, but then a small segment appears again off the coast of the northwestern United States and British

Columbia. What is the origin of this system of ridges, such a prominent feature of the earth? Why isn't it buried under sediments from the continents? And what does it have to do with continental drift and plate tectonics?

The observation usually credited with providing the flash of insight that clarified the origin of the ridge system, and that eventually led to the theory of plate tectonics, came from an unlikely source: the magnetic properties of the seafloor. In their attempt to learn as much about the ocean floor as possible, geophysicists measured, among other things, the local magnetic field over large tracts of seafloor. It was already known that rocks containing magnetic minerals could alter the earth's local magnetic field by small amounts, and on the continents magnetic measurements were being used as prospecting tools. Many economically important mineral deposits contain concentrations of magnetic minerals, and their presence produces characteristic anomalies in the regional magnetic field. Indeed, on the continents the patterns of magnetic field variations are very complex, in keeping with the complicated geology. In contrast, when magnetometers were first dragged behind ships, it was discovered that the magnetic patterns originating in the seafloor rocks are very regular. This observation was first made by scientists from the Scripps Institution of Oceanography, and it puzzled them. They conducted magnetic surveys off the northwestern coast of the United States in the 1950s, and the patterns they mapped were very different from anything that had been seen on the continents. It was eventually concluded that the regular pattern of local magnetic field variations was probably somehow related to the fairly regular seafloor topography of low hills and valleys in this area. But that hypothesis had a fairly short lifetime. In the 1960s, an airborne magnetic survey carried out in the North Atlantic just south of Iceland produced startling and now classic results. In a series of traverses across the axis of the Mid-Atlantic Ridge, scientists from Columbia University's Lamont Geological Observatory found that the seafloor magnetic field varies symmetrically around the exact center of the ridge. Furthermore, they found that the pattern of magnetic field variation was essentially identical on every traverse they made across the ridge, regardless of location. When the data were plotted and contoured on a map of the survey area, they made a zebra-stripe pattern of magnetic inten-

Iceland

10 million
years ago

10 million
years ago

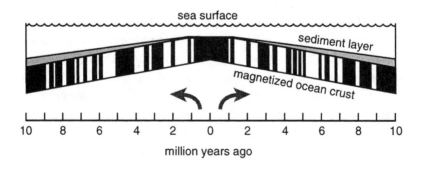

sea surface

sediment layer

magnetized ocean crust

10   8   6   4   2   0   2   4   6   8   10

million years ago

sity variations that was reminiscent of the patterns found by the Scripps scientists in the northeast Pacific, but with the important difference of obvious symmetry (see Figure 5.1). Again, the contrast with typical continental magnetic records was striking. As more data accumulated, it became clear that the same symmetrical pattern occurs *everywhere* along the ocean ridge system.

When igneous rocks cool from their molten state, some of the iron-containing minerals that form in them are magnetized by the earth's magnetic field. It is as though the minerals themselves contain tiny bar magnets—or compass needles—all of which align themselves with the surrounding field. The magnetization is permanent, and thus becomes a fossil record of the characteristics of the earth's magnetic field at the time of rock formation. It is quite stable, and survives over long time periods. The surveys across the Mid-Atlantic Ridge showed that the rocks right at the crest of the ridge are quite strongly magnetized in the direction of today's magnetic field, as might be expected. But the symmetrical, zebra-stripe pattern seemed to indicate that the seafloor is magnetized in strips. Some of these strips, like the ridge-crest rocks, are normally magnetized— they have the characteristics that would be expected for rocks solidifying in the present magnetic field. However, they alternate with strips that are magnetized in precisely the opposite direction, as though the north and south magnetic poles of the earth had been reversed when these segments of the seafloor formed.

The earth's magnetic field is a dipole, meaning that it is similar to the field that would result if there were a giant bar magnet in

FIGURE 5.1   *The magnetic pattern on the seafloor south of Iceland (upper diagram) resembles a series of zebra stripes of alternately normal (black) and reversed (white) magnetization parallel to the Mid-Atlantic Ridge. As basalt wells up and solidifies along the ridge, it is magnetized and then spreads away laterally as shown schematically in the lower diagram. Only the longer polarity intervals shown in the lower diagram are apparent on the seafloor map. The inferred position of the ridge through Iceland is shown in the stippled pattern. Based on Figure 1 of J. R. Heirtzler, X. Le Pichon, and J. C. Barron, in* Deep Sea Research, *v. 13, page 428 (1966). Used with kind permission from Elsevier Science Ltd., The Boulevard, Langford Lane, Kidlington OX5 1GB, U.K.*

the interior. At the time when the first magnetic surveys of the seafloor were being made, most scientists had no reason to believe that the field had been much different in the geologic past than at the present. However, at about the same time, studies of rock magnetization on the continents had turned up a puzzling phenomenon. It was discovered that in some regions where great thicknesses of basalt flows had accumulated, most of the flows were magnetized in the direction of the earth's magnetic field, as expected, but in others the magnetization was reversed. Initially it was believed that some secondary process was responsible, but when very similar sequences of reversed and normally magnetized lava flows were found at several different localities, it was realized that the *earth's magnetic field must have repeatedly reversed polarity over geologic time!* This was a stunning conclusion. Against this background the regular magnetic stripes on the seafloor took on great significance. Although they may not actually have cried "Eureka!," several researchers—Lawrence Morley in Canada, and Fred Vine and Drummond Matthews in the United Kingdom—almost simultaneously realized that magnetic stripes on the seafloor, magnetic polarity reversals, and continental drift were all interconnected. They suddenly understood that the zebra-stripe magnetic pattern on the seafloor records exactly the same sequence of magnetic reversals as the continental basalts.

These observations convinced most geologists that seafloor spreading is a reality. New ocean crust is being formed by lava continually welling up at the center of the ridges. The magnetic pattern is symmetrical because the lava is magnetized as it cools to solid rock and spreads away equally to both sides. The seafloor acts as a kind of gigantic magnetic tape recorder, faithfully recording the reversals of the earth's magnetic field (see Figure 5.1). Because the dates of the various reversals are known from analysis of rocks on land, the magnetic stripes on the ocean floor can be used as time markers. The rate at which new seafloor is being created can be calculated quite simply by measuring the distance from the center of the ridge, where the age of the seafloor is effectively zero, to the various dated reversals. Geologists refer to the magnetic stripes as anomalies and for ease of identification have given them numbers. For those who work with them, they become good friends: "Aha, that looks like anomaly 29R!" (The "R" stands for reversed, as compared to N for normal, the direction of the present-day field.)

Although it varies from place to place, the rate of seafloor creation calculated from the magnetic anomaly data is typically several centimeters per year. This is about the same speed at which your fingernails grow—not very fast, but if you forget to clip them for a while, very noticeable. The continents on either side of the Atlantic are moving apart at this rate, which explains why the oceans are not choked with sediments: they are geologically young. Although a few centimeters per year is indeed slow, the entire Atlantic Ocean can be created in less than two hundred million years at this rate, not very long in geologic terms. In fact, no seafloor in any of the world's oceans is much older than this. In comparison to the continents, the rocks of the ocean floor are mere youngsters.

On either side of the Atlantic, the continents are firmly attached to the rocks of the seafloor. They move apart at a velocity that is governed by the rate of creation of new seafloor at the Mid-Atlantic Ridge. Thus the physicists' objection to Wegener's version of continental drift is not really valid, because the continents don't plow through the rigid rocks of the ocean floor. Instead, continental and ocean crust move together, both part of a single lithospheric plate (see Figures 1.2 and 5.2).

PLATE TECTONICS

The existence of the magnetic patterns of the seafloor, and the understanding of their origin as described above, clinched the case for continental drift, a term that was quickly supplanted by the equally descriptive but more accurate expression "seafloor spreading." The 1960s were a heady time for geologists; the development of the ideas of seafloor spreading and its consequences were called by some a revolution, and likened to the upheaval in physics brought about by the development of relativity and quantum mechanics. The implications of seafloor spreading were rapidly pursued, both by theorists who attempted to describe the process mathematically and by experimentalists who, using increasingly sophisticated instruments, made measurements to test the mathematical theories. Many previously poorly understood

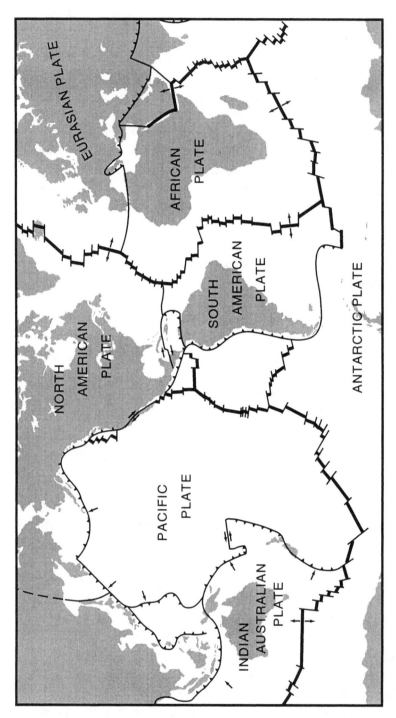

FIGURE 5.2   A map of the world showing the major lithospheric plates. Each plate is bounded by spreading ridges (thick lines), collision/subduction zones (toothed lines), and/or transform faults (thin lines). Names are shown for some of the larger plates, and arrows indicate the relative motion between plates.

phenomena suddenly seemed perfectly normal in the context of seafloor spreading. Before long, seafloor spreading and continental drift had been subsumed into a broader and far-reaching theory, plate tectonics.

What exactly is plate tectonics, and why has it received so much attention in the earth sciences? At the simplest level it is a global framework in which most geologic processes at work today, and throughout a good deal of the earth's history, can be understood. True, there are many details that are not immediately explicable in terms of plate tectonics, but whether this is a shortcoming of the theory or simply our lack of understanding of the processes is not yet clear. However, in broad outline the concept of plate tectonics is a very powerful tool for understanding how the earth works.

The word tectonics is from the Greek *tekton,* for builder or carpenter. The plates in plate tectonics are pieces of the lithosphere, the relatively rigid outer part of the earth that on average extends down to a depth of about 100 kilometers (see Figure 1.2), although in places it may be either thinner or thicker. About ten moderate-sized to large plates, and many more "microplates," are recognized today (Figure 5.2). As explained earlier, it is the lithospheric plates that move around on the earth's surface, not the continents; the continents and oceans are merely passengers, along for the ride. The plates are able to move because the earth's interior is hot, and can deform and flow quite readily. It is difficult to imagine ordinary rocks behaving in this plastic way, but it is useful to remember that other solids we ordinarily consider to be brittle will also flow slowly when subjected to moderate pressures over extended time periods, an example being glacial ice. The bottom of the plates is at a depth where the rocks of the interior are close to their melting point, and friction between the relatively rigid lithosphere and the underlying mantle is near a minimum.

The mechanism for plate movement, the actual driving force, is still not known precisely. But this is no longer cause for derision, as it was in Wegener's day. We know that the plates do move; in fact, using satellites, it is now possible to measure changes in the distance between two locations on different plates accurately enough to prove this, and even to measure the speed of plate motion. We also know that the energy required for plate movement ultimately comes from within the earth, both from its continued cooling from an initially hot state and from heat produced by the radioactive decay of elements such as uranium and thorium that are distributed throughout the interior. This heat is carried toward the surface by slow, solid-state convection, and is eventually lost to space. Coupling between the hot, convecting mantle and the cooler, more rigid lithosphere may be partly responsible for plate movement.

Most geologic action takes place at the boundaries of plates. This includes volcanism, earthquakes, mountain building, metamorphism, and even the formation of many kinds of economically valuable mineral deposits. But not all plate edges are the same. Figure 5.2 shows that in some places plates are moving apart, in others they are colliding, and in a few localities they simply slide past one another. Because there is no fixed frame of reference for examining plate motion, the directions of movement are known only in relative terms. It is possible to stand at a plate edge and determine whether an adjacent plate is moving toward or away from us, but not to know its absolute direction of motion.

The plate boundaries are categorized based on the type of relative movement along them. Each has its own special characteristics; for example, distinctive rock types are produced at the different boundaries. Recognizing these has become particularly important for earth scientists trying to look back into the past, because the ancient equivalents of present-day phenomena can then be identified from the preserved rock record. Once again the usefulness of the principle of uniformitarianism is apparent.

DIVERGING PLATES

Where the plates move apart, there are rifts in the earth's crust. Basalt, the most common product of melting of the earth's interior, wells up to fill them; as we have seen, this is how new seafloor is created. Most of the divergent plate boundaries are found in the oceans. Paradoxical as it may seem at first, the rifts, which are valleys or depressions, are often at the center of ridges, which are broad topographic highs, as shown in Figure 5.3. The ridges exist because of the upwelling mantle material and the heat it carries; as the newly created crust moves away from the ridge it cools, contracts, becomes denser, and sinks to lower elevations. The depth of

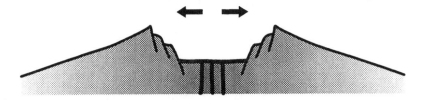

FIGURE 5.3   *A schematic cross section across a midocean ridge, showing a rift valley at its elevated center. The black vertical lines are conduits through which magma flows from the mantle to the seafloor.*

the ocean increases by about a factor of two, from approximately 2½ kilometers to five, from the crest of the ridges to the old parts of the seafloor far removed from the spreading region.

Most of the present-day oceanic ridges actually started out as rifts within a continent. The initial stage in this process is the formation of a deep, steep-walled valley, typically characterized by extensive volcanism. This was the origin of the midcontinental rift that almost split North America apart in the Proterozoic, and the East African Rift Valley is a present-day example. Eventually, as spreading continues, the relatively buoyant continental crust, made from rocks less dense than the basalt erupting in the rift, is split apart. The sea floods in, and a nascent ocean basin is formed. Such must have been the process when the Atlantic Ocean began to open up some 180 million years ago, separating Europe and Africa from the Americas. Today the early stages of continental splitting are occurring in the Red Sea, where Africa is separating from Saudi Arabia along an extension of the Indian Ocean ridge system. All of the world's ocean basins were initiated by rifting, and all are floored by dense basalt. The contrast between the dense crust of the oceans and the lighter, more buoyant crust of the continents is the cause of their difference in elevation.

Along the oceanic ridges, new seafloor is created continually and rolls out symmetrically to either side. While the outline shapes of the continents remain quite recognizable over long stretches of geologic time, the geography of the ocean basins changes much more rapidly. The measured spreading rates along today's ocean ridges range from one or two centimeters per year to as much as twenty. Even at the low end of that range, a 1,000-kilometer-wide ocean basin can be created in 100 million years.

 ## PLATE COLLISIONS AND SUBDUCTION ZONES

If so much new seafloor is being created, and the earth isn't expanding (and there is ample evidence that it isn't), then somewhere on the globe crust must be destroyed to compensate. And indeed that is just what occurs around the margins of much of the Pacific Ocean. Here the lithospheric plates converge, and at these boundaries one of the colliding plates plunges beneath the other and is carried deep into the earth's interior. Such collisional boundaries are known as subduction zones, and they are marked at the surface by both deep ocean trenches and active volcanoes (see Figure 5.4). The spectacular volcanic chains that form the famous ring of fire around the margins of the Pacific—the Andes, the Aleutians, and the volcanoes of Kamchatka, Japan, and the Marianas—all owe their existence to the phenomenon of subduction.

No one is sure just how subduction begins when two plates start to converge, but the key to its operation appears to be density. Dense oceanic crust can be subducted, disappearing into the earth's interior with little trace, while the relatively light continents remain forever on the surface. This is why the ocean floor is young and the continents are old: Seafloor is not only continually created at the ridges, it is also constantly destroyed at the subduction zones. As we have seen, parts of the continents date to almost four billion years, while the oldest seafloor is only about 200 *million* years old. One of the early proponents of continental drift likened the continents to the scum accumulating on a pot of boiling soup, a vivid if less-than-precise analogy.

The reality of subduction is confirmed by the earthquakes that accompany it. Although seismicity exists at all types of plate boundaries, only the subduction zones are characterized by deep earthquakes, some occurring at a depth of 600 kilometers or more. Deep earthquakes were known long before plate tectonics came into vogue. In 1928, the Japanese seismologist K. Wadati reported earthquakes under Japan at depths of several hundred kilometers. Some twenty years later another geophysicist, Hugo Benioff, showed that in other parts of the world as well there are "great faults," characterized by frequent earthquakes, dipping down into the mantle from the ocean trenches. He described these both along the western coast of South America and in the southwest Pacific at the Tonga Trench. These areas were not then recognized as subduction zones; it was only later that it was realized that these great tabular zones of seismicity actually neatly track the path of the plates being thrust down into the mantle (see Figure 5.4). The earthquakes occur because the subducting oceanic plates remain relatively cool as they descend into the hot interior, and, in contrast to the surrounding plastic mantle, are brittle enough even at great depths to sustain the fractures that generate earthquakes. Some of the deepest earthquakes may also occur because the minerals in the subducting plate become unstable under the great pressures to which they are subjected, and break down abruptly to form more compact minerals, with a sudden change in their volume.

In contrast to the relatively quiescent eruptions of basalt at the oceanic spreading centers, volcanism along the subduction zones is often violent. While this activity produces spectacularly beautiful volcanoes, such as Mount Fuji in Japan, it has also caused its share of disasters. Some of the better known include the burial of Pompeii in hot volcanic ash from Mount Vesuvius, the great destruction of life in the eruption of Krakatoa in Indonesia in 1883, and, more recently, the widespread damage done by the eruption of Mount Pinatubo in the Philippines in 1991. Why is there volcanism at subduction zones? The answer was hinted at in Chapter 2: The oceanic plates are wet. There is water in the thick piles of sediments that accumulate on the ocean floor as it moves from its place of creation at the ridge toward its destruction at the subduction zone. In addition, during this long journey some of the minerals of the basaltic crust itself react with seawater to form new minerals containing water. Although part of the sediment gets scraped off and thrust up onto land when plates collide, some of it is carried to considerable depths in the mantle. As these sediments descend along the subduction zone, much of the free water in the pores between grains gets squeezed out by the increased pressure, and makes its way back to the surface. But some remains, as does the water that is tightly bound in the minerals of the crust. Eventually, though, the heat and pressure drive off this water, and it

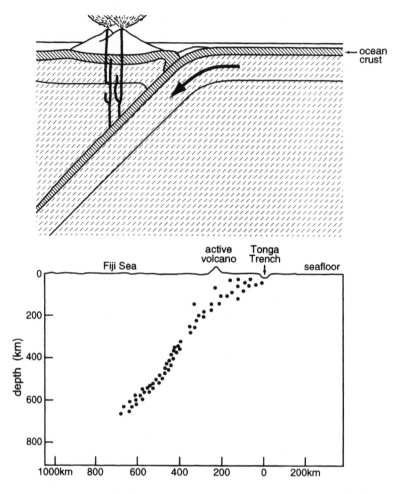

FIGURE 5.4 *The sketch of a subduction zone in cross section (upper diagram, not to scale) shows a lithospheric plate descending into the mantle, with active volcanoes above it. In the lower diagram, the actual positions of recorded earthquakes under the Tonga Trench in the southwestern Pacific are shown as dots. They clearly mark out the location of the subducting plate to about 700 kilometers depth. The horizontal scale marks distance from the trench. Partly based on Figure 4-10 in* The Way the Earth Works, *by P. J. Wyllie. John Wiley & Sons, 1976.*

percolates into the mantle above the subducting plate. It is this process that accounts for the volcanism. At the depths where the water is driven off, the surrounding mantle is already quite hot, and the addition of water lowers the melting temperature sufficiently that it begins to melt. The principle is familiar to dwellers in northern climes, who spread salt on their streets in the winter to lower the melting temperature of ice.

At all of the earth's subduction zones, active volcanism invariably occurs at roughly the same height above the descending plate, about 150 kilometers. This is approximately the depth at which the water-bearing minerals break down, releasing the water that initiates melting. The characteristic rock type in this setting is andesite, named, as you might guess, after a common rock type of the Andes mountains. Laboratory experiments indicate that andesite is just what would be expected if the mantle were to be melted in the presence of water from a subducting plate, and the water also accounts for the violent nature of subduction-zone volcanism. As the magma approaches the surface, the dissolved water and other volatile compounds it contains expand rapidly and explosively in response to the lowered pressure.

Many of the world's largest earthquakes take place along subduction zones. Small wonder when you think about what occurs in these regions: Two gigantic fragments of the earth's surface, each about 100 kilometers thick, are colliding, one plate being thrust beneath the other. Unfortunately, some areas near subduction zones are heavily populated. We can predict with 100 percent certainty that severe, damaging earthquakes will continue to occur in such regions, but that is little comfort in the face of disastrous events such as the Kobe, Japan, earthquake of early 1995.

However, the earth is a dynamic planet, and even subduction zones don't last forever, at least on a geologic timescale. Eventually they cease to operate and another one starts up elsewhere on the globe. What kinds of events could bring the subduction process to a halt? Most commonly, it is collision between continents after the oceanic crust between them is consumed by subduction. Remember that the plates often carry both continental and oceanic crust. While the plate itself may be indifferent to the nature of its passengers, the subduction zone isn't; it simply can't swallow up the low density continental crust. Thus when an ocean basin eventually closes up due to subduction, the two fragments of continental crust simply collide and are welded together, and subduction stops. A simplified sketch of the process is shown in Figure 5.5. It is not quite as uncomplicated as this description might imply; typically collisions between continents involve extensive volcanism, metamorphism, and mountain building, and take a very long time.

The preeminent example of this process from the recent past is the collision between India and Asia to form the Himalayas, described in more detail in Chapter 11. Once, there was a subduction zone near where the Himalayas now stand, dipping under Asia to the north, and a vast ocean between Asia and the continent of India far to the south. The rocks of the Himalayan mountains and Tibetan plateau reveal that this situation lasted for a very long time, and that many small fragments of buoyant crust, transported by the oceanic plate, arrived at the subduction zone and were plastered onto the southern margin of Asia. But gradually the ocean floor was swallowed up, pulling India north-

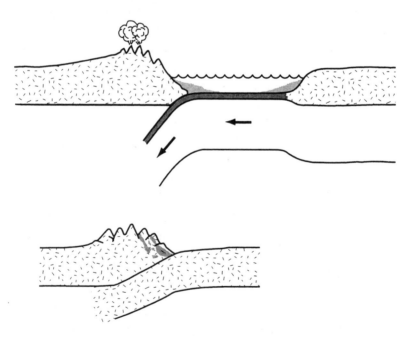

FIGURE 5.5   *A schematic diagram showing how subduction can close up an ocean basin and cause continents to collide, throwing up great mountain ranges such as the Himalayas in the process.*

ward. Between fifty and sixty million years ago, a corner of the continent reached the subduction zone and began to collide with Asia. The momentum of the collision has caused northern India to slide under southern Asia, forming continental crust almost twice as thick as is found anywhere else in the world. Sediments from the precollision margins of the two continents, volcanic islands that existed along their edges, and the rocks of the continents themselves, were caught up in the gigantic crash, folded, faulted, and metamorphosed. The result is the highest mountain range and largest upland area on the earth.

The broad mountainous region of the Himalayas is still considered a plate boundary because there is still relative movement between Asia and India. The region is still being uplifted and is characterized by earthquakes. Indeed, earthquakes relieving stresses in the crust occur today far from the collision zone, particularly in China, as a result of the fact that parts of Asia have been squeezed and rotated to the east as the two plates pushed into one another. Eventually, however, when relative motion between the two formerly separate continents ceases, the Himalayas will be recognized as an inactive suture zone in the interior of a continent. But when this occurs, something else will have to give way to accommodate the new seafloor being created along the ridge that lies far to the south (Figure 5.2). Recent studies of the seafloor near Sri Lanka suggest that a new subduction zone may be in the process of forming to the south of that island nation, which would resolve the geometric puzzle.

Continent-continent collisions like the one that produced the Himalayas appear to have occurred regularly throughout geologic history. Although their high mountains are long gone, such events can be recognized in older rocks from the fact that they typically produce long bands of strongly metamorphosed rocks, all having approximately the same age. A good example is the Grenville Province of eastern North America (Figure 4.3), which was undoubtedly once very much like the present-day Himalayas.

THE SAN ANDREAS FAULT

Like the Himalayas and the midocean ridges, the San Andreas Fault in California is a plate boundary. Los Angeles and San Diego, on the west side of the fault, are on the Pacific plate and are moving in the same direction as the island of Hawaii, while Berkeley, to the east of the fault, moves together with New York and Miami on the North American plate (see Figure 5.6). Plate boundaries that are faults like the San Andreas have been named transform faults, and they occur mostly in the oceans, joining together segments of the spreading ridges. They are the reason the plate edges appear to be so jagged. Along these faults there is no convergence or divergence, the plates simply move past one another. If you tried to invent plate tectonics by breaking up the outer shell of a globe into pieces that subducted at some boundaries and renewed themselves at others, you would find that features resembling transform faults would be a geometrical necessity.

The most famous, or if you like, infamous, transform fault is the San Andreas Fault in California. It too joins segments of the ocean ridge system, but in contrast to most transform faults, it cuts through part of a continent. The evolution of the San Andreas Fault is quite interesting (Figure 5.6). Some fifty or sixty million years ago, there was a subduction zone stretching all along the west coast of North America. Offshore to the west, there was an ocean ridge, where new Pacific seafloor was being created. But the North American plate was moving westward faster than newly minted seafloor could be produced, and eventually the continent

FIGURE 5.6   *From top to bottom, these diagrams show how the western margin of North America evolved as the continent gradually overran the spreading ridge (double lines) in the Pacific. Until well into the Tertiary period, a subduction zone (toothed lines) existed along the entire coast, with Pacific seafloor plunging under North America (top panel). At present (bottom), a transform fault, the San Andreas, joins the remaining segments of the ocean ridge in the Gulf of California and off the coast of the Pacific Northwest. A small sliver of the continent, including Baja California, Los Angeles, and coastal California north to San Francisco, is now part of the Pacific plate, moving northwest relative to the rest of the continent. After Figure 16.24 in* The Dynamic Earth, *3rd edition, by B. J. Skinner and S. C. Porter. John Wiley & Sons, 1995. Used with permission.*

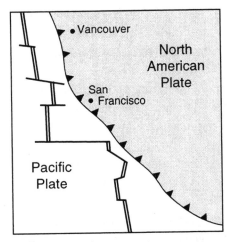

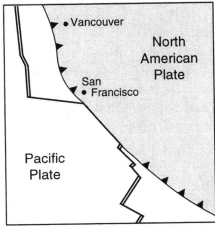

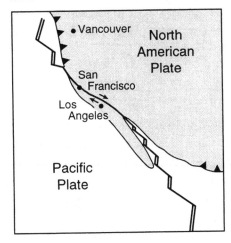

simply overrode the ridge. This first happened about thirty million years ago, and it continued in stages as the small plate between the ridge and the subduction zone was gradually consumed. Some small fragments of it remain, off the coast of Mexico to the south, and off Oregon, Washington, and British Columbia to the north. But as that plate vanished, new plate boundaries emerged to accommodate the global interplay of plate motions. The response of the lithosphere was to crack near the edge of the continent. A small part of North America became attached to the Pacific plate, and the San Andreas Fault was born.

On a global map of the earth's plates such as Figure 5.2, transform faults appear as neat, thin lines. In reality, they are very complex boundaries, particularly when they occur in continental crust. Although there is a single large fault on the geologic map identified as "*the* San Andreas," which is indeed a spectacularly narrow boundary when viewed from the air, the plates actually slide by one another over a very broad region of California characterized by a multitude of faults and deformation features. Many of these more or less parallel the San Andreas itself, and much of California's well-known seismic activity occurs along these lesser-known faults.

To summarize then, the plates that make up the jigsaw puzzle surface of the earth have edges that are spreading ridges, subduction-collision zones, or transform faults, and it is in these regions that most of the earth's volcanism, earthquake activity, and metamorphism occurs. The globe-encircling ocean ridge system, the earth's highest mountains, and its most beautiful and dangerous volcanoes all occur at plate boundaries.

 ## HOT PLUMES IN THE MANTLE

From all that has been said thus far, one would imagine that the interiors of plates are geologically quiescent, and for the most part this is the case. However, there are exceptions. For example, a glance at a map of the Pacific Ocean reveals that there are many islands within the Pacific plate, far from its boundaries. All of them are volcanoes. Many are no longer active, and some are completely overgrown with coral, but they all originated through seafloor volcanism.

How can volcanic activity occur so far from a plate boundary? The Hawaiian Islands provide a very instructive answer. Like many other island groups in the oceans, they form a chain. In fact, if undersea volcanoes are included, it is a very long and impressive chain indeed, extending from Hawaii all the way to the Aleutian Trench (Figure 5.7). In the 1840s the American geologist James Daly observed that the different Hawaiian Islands seem to share a similar geologic evolution, but are progressively more eroded, and therefore probably older, toward the west. Then in 1963, in the early days of the development of plate tectonics, the Canadian geophysicist Tuzo Wilson realized that this age progression could result if the islands were formed on a surface plate moving over a fixed volcanic source in the interior. Wilson surmised that the long chain of volcanoes stretching northwest from Hawaii is simply

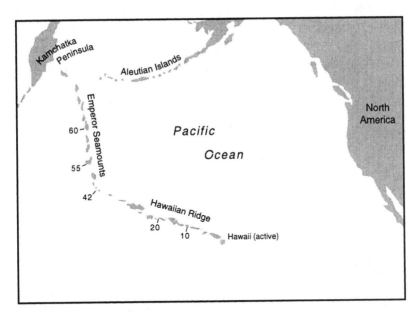

FIGURE 5.7 *A chain of islands and submerged, extinct volcanoes stretches west from Hawaii toward the Aleutian Trench. Dating of rocks from these volcanoes shows that they are progressively older away from the presently active volcanoes of Hawaii (numbers on the diagram indicate ages in millions of years). The sharp bend in the chain reflects a change in the direction of movement of the Pacific plate about 45 million years ago.*

the surface expression of a long-lived, deeply-rooted feature of the mantle.

Although this idea was not immediately accepted, it is now a central part of plate tectonics. An important piece of corroborating evidence is that dating of lavas in the Hawaiian (and other) chains has shown that their ages increase away from the presently active volcano, just as Daly had suggested (see Figure 5.7). Most volcanoes that occur in the interiors of plates are believed to be produced by mantle plumes, fixed sources of volcanic material that rise from deep within the mantle. Their present expressions, such as Hawaii, are referred to as hotspots. Most large, active, within-plate volcanoes have a hotspot trail of progressively older extinct volcanoes that mark out the path of the surface plate over the deep-seated plume. The plumes apparently originate at great depths, perhaps as deep as the boundary between the core and the mantle, and many have been active for a very long time. The oldest volcanoes in the Hawaiian hotspot trail have ages close to eighty million years. Tahiti and Easter Island in the Pacific, Réunion and Mauritius in the Indian Ocean, and indeed most of the large islands in the world's oceans, owe their existence to mantle plumes.

 ## HOW LONG HAS PLATE TECTONICS OPERATED?

In addition to the fact that many of them are nice places to visit, the oceanic volcanic islands and their hotspot trails are especially useful for geologists, because they record the past locations of the plate over a fixed source. They therefore allow the process of seafloor spreading to be run backward, permitting reconstruction of the geography of continents and ocean basins in the past. Because the plates are rigid, the position of the Pacific plate fifty million years ago can be determined by moving it such that a fifty-million-year-old volcano in the hotspot trail sits at the location of Hawaii today.

However, because the ocean basins really are ephemeral features on geologic time scales, reconstructing the world's geography by backtracking along the hotspot tracks works only for the last 5 percent or so of geologic time. The same problem attends efforts to trace the history of seafloor spreading by using the seafloor magnetic reversal patterns. How can we obtain information about the operation of plate tectonics at earlier times? Beyond about 200 million years, the age of the oldest seafloor, the only available clues are from the continents, and they are much more difficult to find and decipher. For example, the magnetic properties of continental rocks can sometimes be used to get information about their position relative to the magnetic pole when they formed, but this can be done only if the rocks occur today in exactly the same orientation they had when they acquired their magnetic characteristics. If they have been folded or tilted, the interpretation is much more difficult, if not impossible. Furthermore, because the continents have wandered over the entire globe throughout geologic history, for very ancient rocks it may not even be possible to determine whether the magnetization occurred in the Northern or the Southern Hemisphere.

Fossils sometimes also provide information about past plate positions. Wegener's arguments for continental drift rested partly on fossil evidence suggesting that some continents, now widely separated, were once joined together. They may also indicate latitude, or at least they can be used to distinguish tropical from temperate or polar environments. However, the fossil record only characterizes the more recent parts of geologic history, and is not useful in the Precambrian. For the Proterozoic and Archean eons, the relative positions of the plates, or in some cases even what constituted the plates, is poorly known. Indeed, there has been heated debate about whether or not plate tectonics even operated in the distant past. There is, however, an abundant record of continental sutures in the Precambrian, as noted in Chapter 4, and these must mark the sites of ancient subduction zones where continents or continental fragments collided with one another as ocean basins closed up. The character of the rocks in these zones is generally similar to that observed in more recent examples. A telling clue in many of these suture zones is the presence of small slivers of ocean floor that were thrust up onto the continent during the collision, a clear indication that they originated at a convergent plate margin where seafloor was being subducted. Thus, although there are some skeptics, most geologists are convinced that plate tectonics has been operating roughly as it does today for billions of years, and perhaps even from the very beginning of earth history.

6

NATURE'S TIMEPIECES

MUCH HAS BEEN said about time in previous chapters. Geology is essentially a historical science, and time is therefore of central importance. The earth formed 4½ billion years ago, the Atlantic Ocean began to open up about 200 million years ago, the dinosaurs became extinct 66 million years ago. All of these statements give quite precise dates to important events in the history of the earth. How can we be sure they are correct?

The early Greeks and Romans had deduced from their observations of nature that sedimentary rocks represented long spans of time. However, it was James Hutton, that remarkable Scottish geologist with his ideas about uniformitarianism, who first began to convince people in modern times that the rock record is truly ancient. His approach was simple and classically scientific: He observed the processes of sedimentation taking place around him, and realized that they are generally very slow. He deduced that the thick outcrops of now-consolidated sediments that he saw in cliff faces must therefore represent very long periods of sediment deposition. Darwin, who was well acquainted with Hutton's ideas, also recognized that vast lengths of time would be required to account for the processes of biological evolution that are recorded in the fossil record.

Neither of these men, nor any of their contemporaries who were convinced of the great age of the earth and the slow pace of geologic change, had a way to determine geologic time accurately. Nevertheless, timescales of hundreds of millions of years were estimated, figures that were revolutionary in their time. Many of the influential elite of the day had an education based in theology, and such ideas were in direct opposition to the literal interpretation of the Bible. In fact, it was because of opposition from the Christian church that the ideas of the early Greeks about the great age of sediments and fossils had been discarded. Furthermore, like Wegener's hypothesis of continental drift, the concept of an earth this ancient was attacked by other scientists. Especially influential was the British physicist Lord Kelvin, who argued in the latter part of the nineteenth century that the earth could not be more than forty million years old and was probably closer to twenty million, based on his calculations of its cooling history. His arguments seemed to be correct, and were difficult for geologists to counter in a quantitative way—but they were contradicted by the geologic evidence.

One of the flaws in Lord Kelvin's argument, we know now, is that the earth contains a number of naturally occurring radioactive isotopes. These decay slowly, releasing heat in the process and effectively prolonging the cooling down of the earth. However, radioactivity was unknown when Kelvin made his calculations, so he was unable to account for its effect.

There is an amusing story about Ernest Rutherford, one of the early pioneers in research on radioactivity, that is related to Lord Kelvin's estimate of the age of the earth. Rutherford was giving a lecture about the heat produced by radioactive decay, but was nervous because Lord Kelvin, still a powerful force in British science, was in the audience. In a smooth twist, he announced during his talk that Kelvin had in fact anticipated the discovery of radioactivity because his calculation of the earth's age had been made with the proviso that the result would be different if a new source of internal heat were to be found. It is said that Lord Kelvin, then eighty, had nodded off during the lecture but awoke with a broad smile when he heard Rutherford's pronouncement.

In addition to producing heat within the earth, radioactivity also provides geologists with a whole series of reliable "clocks" for measuring ages of rocks and the rates of various geologic processes. But before discussing the details of how this is done, it is worth examining the way in which the dimension of time was approached *before* the advent of radioactive dating. In fact, most details of the last 550 million years or so of the geologic timescale in Figure 1.1, the whole of the Phanerozoic eon, had been worked out long before the actual dates of the boundaries were determined. The relative positions of the various divisions were known, but their durations were not.

The concept of relative time is a simple but very powerful tool for determining the age relationships among different rock units, as noted in Chapter 4 and illustrated in Figure 4.1. The approach is straightforward, and often boils down to the simple question, Is A older than B, or vice versa? One of the most obvious aspects of relative time must have been known intuitively for millennia, but was only stated formally in the seventeenth century: In a sequence of sedimentary layers, the youngest material is at the top. The man who devised this law was a Danish anatomist who lived in Italy and latinized his Scandinavian name (Niels Stenson) to Nicolaus Steno. Steno made important contributions in medicine as well as in geology and mineralogy, but, sadly, he became a priest at the age of thirty-seven and abandoned science. By pointing out the obvious—that sediments deposited in water must form horizontal layers initially, regardless of their present orientation, and that the youngest layers must be at the top—he laid the groundwork for the geologic timescale.

However, the earth is a dynamic place, and it is not possible to find at any single locality a complete, layer-upon-layer, sedimentary record of the whole of the Phanerozoic. How then is it possible to construct the geologic timescale, even in terms of relative time? The answer lies in evolution, and the continuously changing nature of the assemblages of fossils preserved in sedimentary rocks. Indeed, more than half a century before Darwin had published his ideas about evolution, an English engineer named William Smith, who was making maps along the canals of southern England, discovered that he could draw up a combined vertical sequence of the sedimentary layers that he found at different elevations in different localities. He did this using fossils, or more accurately the assemblages of fossils that occurred in the different sedimentary rocks he was mapping. He was able to put together a composite sequence because in many places there was partial overlap. This is easy to envision if you consider specific groups of fossils in sedimentary rocks to be characterized by letters of the alphabet, with A being the oldest (see Figure 6.1). A cliff at one locality may reveal sedimentary layers with the fossil groups A, B, C, and D, at another place groups C, D, and E may be present, and at yet another groups C, E, F, and G are found. By matching

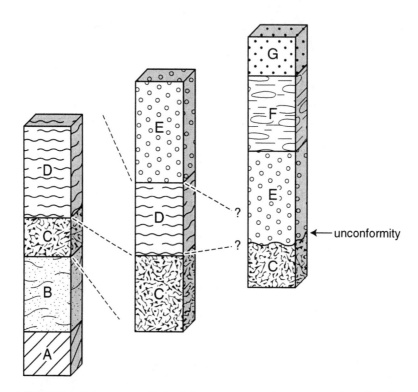

FIGURE 6.1 *As explained in the text, sedimentary rock units often contain diagnostic assemblages of fossils (here indicated by letters of the alphabet) that allow correlation between different localities (sometimes correlation is possible just based on the rock types, but fossils are more reliable). From such correlations a time sequence can be built up; for example, in this simple illustration it is clear that fossil groups A and B are older than F and G, even though they never occur at the same locality. Note that sometimes rock units disappear altogether, as happens with unit D. In the rightmost column there is an* unconformity *between C and E, indicating that there is a time gap in the record. In this locality, unit D and part of C were eroded away before the deposition of unit E.*

the common fossil groups at different locations, an entire vertical sequence can be constructed, just as if the sediments were all present at one place. And by Steno's rule, the oldest rocks are at the bottom, the youngest at the top. In this simple example, although groups A and B are never found at the same localities as groups F and G, it is obvious that F and G are younger, based on the composite, relative time sequence. In principle, we also know that if any one of these fossil groups is encountered anywhere else in the world, it can be placed in its proper place in the timescale of evolution, relative to the other groups.

This is essentially the procedure that was used to build up the relative time scale of Figure 1.1, without the actual ages. Of course, it was not quite as simple as might be imagined from this example. In spite of the fact that the timescale includes data from far-flung localities, there are some parts of the geologic record that are very poorly represented in sediments anywhere on the globe. In fact, because of plate tectonics, most of the sedimentary record laid down in the oceans in the past has been destroyed—either dragged down subduction zones, or metamorphosed beyond recognition in collisions between continents. And for sediments that have survived, usually those that were deposited along the margins of continents or in shallow inland seas, there are geographic differences in the fossil record that must be taken into account, just as today there are marked differences in the flora and fauna of, say, the coral reefs surrounding a tropical Pacific island compared with the Atlantic waters around Iceland. But the relentless beat of evolution, and the similarity, if not exact identity, of many species across geographic boundaries at a given point in time, have made this approach remarkably successful.

Relative dating using fossils permitted early geologists to work out the sequence of major events that occurred during the Phanerozoic eon. They knew, for example, that there were fish on the earth before there were dinosaurs or mammals. They were able to determine that the extensive coal deposits in eastern North America and western Europe formed in ancient swamps long before the chalky sediments that now make up the White Cliffs of Dover formed on the ocean floor. But for rocks without fossils, especially those of the Precambrian, they were adrift. There was a general sense that the most highly metamorphosed rocks were probably older than those that were less deformed and changed, but there was no way to determine whether such rocks in India were older than similar-looking ones in Canada, or vice versa. There was also no clue that the relative timescale they were developing for the Phanerozoic actually involved only about twelve percent of geologic time. And even where the relative sequence was reasonably well known, there was no way to estimate the duration of the various parts of the scale. That capability came only later, for the most part in the second half of the twentieth century, and it is still being refined.

 ## DATING WITH RADIOACTIVITY

The timekeeper *par excellence* for geologists is radioactivity. Fortunately, there are many naturally occurring radioactive isotopes with properties that make them useful for geologic chronology. Their importance cannot be overemphasized. It is only because of them that it is possible to construct the history of the earth related in this book.

How can radioactivity be used to determine ages or timescales? The subject is a complicated and highly technical one; thousands of scientific papers and many books have been written on the topic. It will only be possible here to provide a brief outline, with some examples. But the basic premise is actually quite simple: Radioactive isotopes decay at a constant rate. In this sense they are exactly analogous to ordinary clocks. We know that in every minute a clock ticks off sixty seconds; we also know that in any sample that contains uranium, about 1½ percent of its uranium 238 atoms will decay to lead every 100 million years. By measuring the amount of uranium that has decayed away over the lifetime of a particular sample (or, alternatively, the amount of lead that has been produced by the decay) its age can be determined.

Most chemical elements in the periodic table have several isotopes. As already described in Chapter 2, all isotopes of an element have essentially the same chemical behavior. Each has the same number of protons in its nucleus, and the same number of electrons surrounding the nucleus; however, each has a slightly different weight, because each contains a different number of neutrons. Isotopes are identified by a number that refers to the sum of the protons plus neutrons in the nucleus (and therefore its weight); thus in every mouthful of air you breath most of the oxygen atoms are oxygen 16, but some are oxygen 18 and an even smaller number are oxygen 17. As far as your body is concerned, however, it's all oxygen.

Radioactive isotopes are unstable. Radioactive decay works toward stability by changing the balance of neutrons and protons in the nucleus. This occurs by the ejection of particles from the nucleus at high energies, and in the process, a different chemical element is formed—for example, we have seen that uranium decays to lead (although in this particular case, the transformation involves a whole series of radioactive decays, not just a single step). Radioactivity was discovered during the last few years of the nineteenth century, and has been studied intensively ever since. It was quickly learned by experiment that radioactivity is a statistical phenomenon, that is, each radioactive isotope is characterized by a probability that it will decay in a given period of time. This is most easily visualized by imagining a large number of radioactive atoms in a beaker. If observed over some time period, say one minute, a certain fraction of the atoms will decay; if observed for a second minute, the same fraction of the remaining atoms will decay, and so on. Because it is a statistical phenomenon, and especially if the number of atoms in the beaker initially was small, the fraction that decays may vary slightly from minute to minute, but on average it will be constant. The same experiment conducted at different times, and under a wide range of environmental conditions, would give the same result. This indicates that the probability for decay of a given isotope is a constant. The decay constant is perhaps most easily thought of in terms of the concept of half-life, which is the time period required for one-half of the atoms initially present in a sample to decay. Mathematically the half-life is directly related to the decay constant, and for most radioactive isotopes, it has been measured quite accurately. It is this knowledge that is the key to all of the "absolute" dating methods used in geology.

You may have recognized from this description that radioactive decay is exponential, that is, the actual number of atoms that decay initially is large, but becomes smaller with time. It is the fraction that decays in each time period that remains the same, as depicted in Figure 6.2.

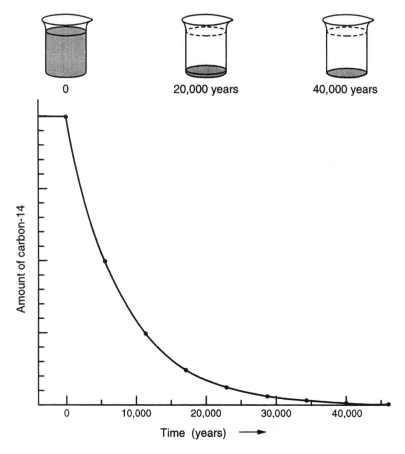

FIGURE 6.2   *The amount of the radioactive isotope carbon 14 (shown here in arbitrary units), in, for example, a plant remains constant as long as it is alive and exchanging CO_2 with the atmosphere. After it dies (indicated by time zero on the graph), its carbon 14 content decreases by one-half every 5,700 years, decaying to nonradioactive nitrogen. The dots on the curve in this diagram are spaced one half-life apart, and it is obvious that after five or six half-lives, very little carbon 14 remains. The same process is shown in the upper part of the diagram by the disappearance of carbon 14 from an initially full beaker.*

There are actually quite a few radioactive isotopes that occur naturally, more than most people realize. There are also many in the environment that have been produced by man in nuclear bombs and reactors. Some of these have been used in geologic studies, but they will not be considered in this chapter.

Why do unstable radioactive isotopes exist in nature? Together with the stable elements, most were produced by nuclear reactions in the interiors of stars or in the supernovae explosions that occur regularly in our galaxy. They were part of the material incorporated into the earth when it formed, and those with very long half-lives have only partly decayed away. These are still found on the earth. But there are others with half-lives so short that any amount present when the earth formed would long since have decayed away. The fact that these exist means that they must be produced in some other, ongoing, process.

A good example from the latter category is carbon 14, an isotope that is familiar to many people in connection with carbon dating. It has a short (geologically speaking) half-life of about 5,700 years, so that none of the carbon 14 now observed can be left over from the time of earth formation (as should be obvious from Figure 6.2). Instead, our planet's supply of this isotope is constantly replenished by nuclear reactions in the atmosphere. This is a fortunate circumstance for archaeologists and climatologists, who make extensive use of carbon 14 for dating.

The nuclear reactions that produce carbon 14 in the atmosphere are caused by cosmic rays, the name given to the particles—mostly individual atoms—that permeate space and frequently impinge on the earth. Many of these atoms come from our own sun, accelerated into space in large numbers when great tongues of flame, solar flares, spurt millions of miles above the sun's surface. Some, with even greater energies, are travelers from regions far outside the solar system. But regardless of their origin, when high-velocity cosmic ray particles collide with the atoms of the earth's atmosphere, nuclear reactions occur just as they do in man-made particle accelerators. A by-product of many of these reactions is neutrons, and, when a cosmic-ray-produced neutron hits and is captured by the nucleus of a stable nitrogen 14 atom (nitrogen being the most abundant element in the atmosphere), knocking out a proton in the process, radioactive carbon 14 is made.

Most of the carbon in the earth's atmosphere is combined with oxygen, forming the molecule carbon dioxide. This is also the fate of the radioactive carbon 14 atoms produced by cosmic rays, so that in any particular sample of carbon dioxide from the atmosphere, a fixed fraction of the carbon atoms is carbon 14. Because the carbon in living things ultimately comes from the atmosphere, via photosynthesis in plants, it too contains the same fixed fraction of its carbon as carbon 14. This is the basis for the use of this radioactive isotope as a chronometer.

Carbon 14 dating has been used to determine the age of the Shroud of Turin, to date shells from North American Indian middens, and to find the age of prehistoric volcanic eruptions in the Hawaiian Islands. How then are actual samples dated using this technique? First, material has to be found that incorporated carbon dioxide (and thereby carbon 14) from the atmosphere. Anything that contains carbon and was alive at the time of the event being dated will do, although some substances are more useful than others. Preserved plant material such as wood, or even charcoal, is often used. When the plant died, or the tree was cut down or burned or engulfed in a lava flow, it ceased incorporating carbon from the atmosphere, and from that time onward the radioactive carbon 14 it contained decayed away according to its well-known decay constant, exactly as portrayed in Figure 6.2. If an old wood sample were exactly one half-life of carbon 14 old—5,700 years—it would contain exactly 50 percent of the carbon 14 found in present-day plants, if it were two half-lives old, 25 percent, and so on. From Figure 6.2 it is obvious that there is not much of the radioactive isotope left after several half-lives. Yet with modern techniques, extremely small amounts of carbon 14 can be detected, and ages up to forty or fifty thousand years can be measured. That is more than eight half-lives, and therefore less than $\frac{1}{256}$th of the original carbon 14 remains in a sample of this age.

The only uncertainty in this treatment concerns the amount of carbon 14 in the ancient atmosphere: It may have been different from today. However, there are various ways to check this possibility—for example, by calibrating the carbon 14 ages against some other dating technique. Although small fluctuations have been noted, such tests have shown that in general the assumption of approximately constant carbon 14 in the atmosphere works well over the timescale that can be investigated using this method.

This brief description provides an example of how radioactive isotopes can be used to measure ages of objects or events. However, the half-life of carbon 14 is so short that it is only useful for the chronology of the very recent past. For the rest of the geologic timescale, much longer-lived radioactive isotopes are employed, and the ways in which they are used are somewhat different.

In Chapter 2 we mentioned lead isotopes, and their utility for measuring the age of the earth and dating the weathering-resistant mineral zircon. We saw that various isotopes of lead are the stable end products, usually called daughters, of the decay of radioactive uranium and thorium. The uranium-lead method was actually the first technique ever employed to determine the ages of rocks using radioactive decay, and it is still one of the most useful in geology. Other parent-daughter isotope pairs that are in common use involve the decay of an isotope of potassium to an isotope of the gas argon, and the decay of rubidium 87 to strontium 87. The parent isotope in each of these cases is a widespread constituent of rocks in the earth's crust, and has a half-life that is long enough for the method to be useful over the whole span of earth history.

In principle, the methods employing long-lived radioactive isotopes are similar to the carbon 14 method, but there are some important differences. One is that the parent isotopes are not continuously produced in the earth, but are gradually decreasing in abundance through radioactive decay. Thus there is much less uranium in the earth today than when it was formed; much of it has decayed to lead.

For most of the commonly used dating methods, the procedure followed is to measure the amount of the daughter isotope that has been produced over time rather than the amount of the radioactive parent isotope left in a sample, as is done for carbon 14. This obviates the need to know the amount of the parent isotope that was present when the radioactive clock started. Because each parent atom decays to a daughter atom, the number of daughter atoms is always equal to the number of parent atoms that have decayed.

The potassium-argon dating method is a good example of how this works. Potassium 40 is the only one of the three potassium isotopes occurring in nature that is radioactive. Although potassium 40 is not very abundant, constituting only about 0.01 percent of the element, potassium is common in minerals in the earth's crust, and thus many kinds of rocks can be dated using this technique. The half-life of potassium 40 is 1.3 billion years, making it useful for dating rocks as old as the earth and as young as

100,000 years, or even less. The daughter isotope of this decay is argon 40, a gas, and although argon is not a rare element—it makes up about one percent of the atmosphere—most igneous rocks, especially volcanic rocks that are erupted at the earth's surface, contain no argon 40 whatsoever when they form. Any argon that was dissolved in the molten lava simply degasses into the atmosphere during eruption. Thus all of the argon 40 measured in an old volcanic rock must have come from radioactive decay of potassium 40 over the lifetime of the sample. Because the half-life is well known, it is a straightforward matter to calculate the time required for this amount of argon to accumulate. Some common minerals, such as feldspar and mica, are rich in potassium and therefore make especially sensitive chronometers.

Other long-lived radioactive isotopes used for geochronology are employed in analogous ways, although each has its own peculiarities. Because these methods involve different chemical elements, some are better than others for dating particular types of rock. However, it is often the case that the same rock can be dated by more than one technique. Although the radioactive isotopes involved may have quite different half-lives, and the parent and daughter elements quite different chemical properties, the same age is generally found. This provides considerable confidence that the dating methods are sound, and also that the relevant half-lives are accurately known.

But what is it, exactly, that is being dated? The example given above of a volcanic rock accumulating argon 40 since the time of its eruption is straightforward: The time of the eruption, which is the same as the age of the volcanic rock, is determined. But what about sedimentary or metamorphic rocks? Do they follow the same rules? The answer is: Yes and no.

Take the case of a sedimentary rock. Suppose a potassium-rich mineral is separated for dating by the potassium-argon method, and an age of 300 million years is measured. Is this the time when the sediment was deposited? In general the answer is no, because many of the minerals in sediments are fragments of preexisting rocks. They have been carried from their original source to the site of deposition by rivers and ocean currents. The measured age for the potassium-rich mineral grain is likely to be correct, but it probably reflects the time of formation of the granite from which the grain was weathered, not the time of its deposition as part of a sedimentary rock. All that can be said is that the sedimentary rock can't be older than 300 million years. It must be younger than its constituents; how much younger is not always easy to determine.

The case for metamorphic rocks can be even more complicated. All of the dating techniques are sensitive to temperature to some extent, especially the potassium-argon method. Heat up a potassium-rich mineral, and some of the argon 40 gas that has been accumulating is likely to diffuse away into the atmosphere. Because metamorphism invariably involves elevated temperatures, most rocks lose some of their argon during metamorphism. If the loss is complete, the radioactive clock is reset to zero, and the age that is measured is the time of metamorphism. But generally the loss is partial, and furthermore it is usually not possible to determine just how much argon was lost. Sometimes this can be sorted out by using several different techniques, or by analyzing a variety of minerals with different temperature sensitivities. However, the information is not always easy to interpret. Still, great advances have been made in recent years in understanding the behavior of elements such as argon in various minerals during heating, and in some cases it may even be possible to reconstruct the temperature history of a rock from careful analysis of its isotopic composition. This approach has been especially fruitful for investigating the history of mountain ranges such as the Himalayas, where deeply buried (and therefore very hot) rocks have been uplifted to cooler regions near the surface, where they begin to retain argon. In favorable cases, the chronology of the uplift can be determined quite accurately.

INDIRECT DATING METHODS

In many circumstances it may be possible to determine the age of a rock without measuring parent and daughter isotopes directly. This is often especially useful for sediments, which, as discussed above, are frequently not amenable to direct age measurement. Sediments can often be dated indirectly, sometimes quite precisely, using fossils.

Fossils are the preserved remnants of living organisms. Sometimes they are just the imprints of soft tissue, now decayed away, as in the case of many plant fossils. More often they are the hard parts of organisms—shells, teeth, bones. Unfortunately, none of these usually incorporates the radioactive isotopes used for dating in large abundance, and at any rate the chemical compositions of fossils are often completely changed by circulating water long after their deposition, with little effect on physical appearance. However, because life evolves, fossils are natural chronometers, their morphology and other characteristics changing with time. If the time span over which a particular organism or group of organisms lived on earth can be determined, then its appearance as a fossil automatically dates the rock in which it occurs.

Fortunately, it has been possible to assign quite precise ages to most fossil organisms, because there are some components of sediments that can be dated accurately, even if the fossils themselves can't. For example, explosive volcanic eruptions produce large clouds of ash that settle out over geologically short intervals as layers of gritty rock and mineral fragments, as many residents of the state of Washington learned to their dismay after the eruption of Mount Saint Helens in 1980. Geologists, however, are partial to ash layers because they are marker beds in otherwise unremarkable sediments. The minerals they contain can often be dated by methods such as the potassium-argon technique, providing accurate ages at intervals throughout the sediment column. In this way, using ash layers and other datable components in the sediments, a time framework for the fossil record has been built up through thousands of analyses worldwide. In fact, this is exactly how the dates have been assigned to the geologic timescale of Figure 1.1, which existed as a relative sequence even before radioactivity had been discovered. The precise ages of some of the boundaries in the timescale are being further refined even today by careful, detailed studies that combine analysis of fossil assemblages with precise dating of sedimentary components such as ash layers.

Yet another indirect dating method is based on the periodic polarity reversals of the earth's magnetic field that were discussed in the previous chapter. By analyzing basalt flows on the continents for both their ages (using the radioactive isotope methods just described) and their magnetic characteristics, a chronology of reversals has been constructed that is now quite detailed. It is so detailed, in fact, that the ages of various parts of the seafloor can be determined simply by matching the zebra-stripe reversal pattern on the seafloor to the dated sequences from the continents.

Magnetic chronology is not confined to igneous rocks, however. Like the basalts of the seafloor, sediments also contain magnetic minerals, and as they slowly settle to the ocean floor they too align themselves with the prevailing magnetic field. Thus the earth's magnetic polarity reversals are also recorded in sediments, and again, by cross-correlating with the known chronology, the age of the sediment can be determined from the magnetic pattern.

Geologists have expended tremendous efforts to determine and refine the timescale of the earth's history. The best age determinations for ancient, Precambrian rocks are uncertain by less than a percent, which means that events that occurred only a few million years apart can be put in proper sequence even in rocks that are three billion years old. That is truly a remarkable feat. It is equivalent to putting into their correct sequence events that occurred just a few hours apart a year ago, by somehow measuring the consequences of those events today. It is also remarkable that we know quite accurately the pace of evolution, the exact date of the extinction of the dinosaurs, and the timing of continental splitting and collision. All of this knowledge has been gleaned using the radioactive clocks discussed in this chapter. Throughout this book, whenever ages are mentioned, they are ultimately based on these same geologic timekeepers.

7

THE CAMBRIAN
EXPLOSION

AFTER THE DIVERSION of the last two chapters to delve into the subjects of plate tectonics and geologic time, the story of the earth's history picks up again where we left off: at the end of the Proterozoic eon. The next major division of geologic time (see Figure 1.1 again) is the Paleozoic era, which began with the Cambrian period about 540 million years ago. For the record, it should be noted that there is some uncertainty about the exact age of the Proterozoic-Cambrian boundary. Even recent estimates range between approximately 530 and 600 million years. Such variability is part of the natural progress of science, not, as some would have it, a sign of weakness of the approach. The reasons for the uncertainty lie both with the technical problems of dating the rocks, and, because not all rocks can be dated, with finding appropriate samples that are at or very close to the boundary. There is also the problem of deciding precisely where to place the physical boundary in a particular sequence of sedimentary rocks. The value of 540 million years used here is based on careful uranium-lead dating of zircon crystals extracted from a layer of volcanic ash found in sediments from Yunnan, China. There is no question that the age for the ash layer is accurate. The uncertainty has to do with where it lies relative to the actual boundary. Paleontologists have inferred from their fossil contents that the sediments immediately above and below the ash layer were deposited very close to the beginning of the Cambrian. Regardless of the precise date, however, from the beginning of the Cambrian onward, geologic history is inextricably tied to the history of life on earth, a tale that is told by the record of fossils in rocks.

TNT was unknown 540 million years ago, and the explosion that geologists refer to was not of the violent kind often implied by that word. Rather it describes the very rapid proliferation of a truly amazing diversity of living things on the earth. Most of these creatures are now extinct, and we know about them only from their fossils.

 ## THE FOSSIL RECORD

The fossils used by geologists to trace evolutionary pathways and deduce past climates come in a variety of forms. Some are almost unchanged from their original state, such as the skeletons of saber-toothed tigers recovered from the La Brea Tar Pits in Los Angeles, but most have been changed by chemical reactions while retaining their outward appearances. The most commonly preserved fossils are the hard parts of animals that are constructed of common minerals—bones or teeth made of phosphate minerals; shells made of calcium carbonate. Soft tissue usually decomposes too rapidly to leave much of a record, although in a few sedimentary environments it is preserved. This is fortunate, because most Precambrian and early Cambrian animals were soft-bodied, and their fossils have been critical for understanding the nature of the Cambrian explosion.

The chemical reactions that often drastically alter the mineralogical makeup and chemical composition of fossils, hard parts and soft tissue alike, usually leave the morphology and internal structure of the plant or animal untouched. The reactions typically occur after burial of the organism in sediments, when circulating water, carrying dissolved minerals, reacts with the original material, transforming it. A good example is petrified wood, which retains its original characteristics in detail, tree rings and all, in spite of the fact that the cellulose and other components of the original trees have been completely replaced by silica, the same compound that constitutes the common mineral quartz. As its name implies, the wood has been turned to stone.

Some very useful fossils are not remains of organisms at all, only their traces: the burrows of worms, scratchings of crabs, or footprints of dinosaurs. Like an expert tracker who can tell the sex, height, and weight of a person from indistinct footprints, some paleontologists have gleaned tremendous amounts of information about ancient organisms and their behavior from these trace fossils. Their job is even more difficult than the tracker's, because in many cases it's not at all clear what kind of animal made the trace in the first place.

As a geology student, I was not enamored of paleontology. I remember spending long winter Friday afternoons in an overheated room of an ancient building sketching fossils of ancient creatures under the tutelage of a kindly but also ancient instructor. There was little emphasis on behavior, or evolution, or process, only classification. I thought that geophysics and geochemistry were much more interesting; besides I was never very satisfied with my drawings. But if you reflect for a moment, I think you would agree that with the right approach there could hardly be anything more exciting than holding in your hand a 500-million-year-old rock, scraping out from it the fossil of some unknown creature, and attempting to reconstruct the world of half a billion years ago. Paleontologists have done just that. For a myriad of fossilized animals from the early Cambrian, they have been able to work out modes of locomotion, diet, and place in the greater scheme of evolution.

In many parts of the world it is hard to avoid seeing fossils if you are even marginally observant. Picking up pebbles along a beach, a stroll in the countryside, or even a visit to the bank—if the building is made of sedimentary rock such as limestone—can bring you into contact with fossils. The abundant record of fossils begins with the Cambrian period, and for a long time it was thought that this was so only because it was at this juncture in the earth's history that animals developed hard parts. But it is now recognized that this is only part of the story. There are some rock units of Cambrian age which are chock-full of fossil animals that had no shells, no bones, no teeth—they were composed entirely of soft tissue. Such material is usually quickly destroyed, but in some circumstances, for example, if it is buried rapidly in an environment with little free oxygen, even soft tissue can be fossilized. And if special geologic conditions preserved such fossils in the

Cambrian period and later, why not earlier as well? The answer, it appears, is that there was little to preserve. The diversity of multi-cellular animals underwent a huge expansion beginning near 540 million years ago, and it is this sudden change in the fossil record that defines the boundary between the Proterozoic and Cambrian. As the chapter title indicates, it has been called the Cambrian explosion, and explosion it was. Some researchers have estimated that as many as 100 phyla (the broad divisions of the animal king-dom, based mainly on body structure) developed in the Cambrian; in contrast there are only about thirty in existence today. Whether or not this very large number of early phyla is borne out by future studies, it is an incontrovertible fact that the end of the Proterozoic was marked by a radical change in life on the earth.

In Chapter 3 it was noted that the fossil record of life actually starts long before the Cambrian, in rocks that are 3.5 billion years old, and that there are indications, although no unequivocal fossils, of even more ancient life. However, for a very long stretch of geo-logic history after this early appearance—more than two billion years—the evidence suggests that only the simplest, single-celled organisms inhabited the earth. More complex and mobile animals leave burrows and tracks in the mud even if their soft bodies are not preserved. Such features are common in sediments of Cam-brian and younger age, but, although they have been sought dili-gently, are rare before this and none have been found in rocks older than about one billion years.

Very late in the Proterozoic, but before the Cambrian explosion proper, a number of soft-bodied animals appeared in the oceans. This group of organisms is known among geologists as the Edi-acaran fauna, after Ediacara, Australia, where the fossils were first found. The time when the Ediacara fauna enters the fossil record has not been determined very precisely, but it is probably less than 100 million years before the beginning of the Cambrian period. In the conventional wisdom, it contains the precursors of some Cambrian and even modern animals. But some recent research suggests otherwise. The body plan of these animals is, in one view, in a category of its own, distinct from those of any modern or Cambrian organisms. The fossils show that the Ediacaran animals were basically flat creatures, with many sections (they have been described as "quilted") which lay about on the seafloor like so many minute carpets. Apparently they had no internal structures, and some paleontologists have proposed that they represent an entirely separate kingdom of animals from any that we know today. If this interpretation is correct, then the Ediacaran animals are a dead-end branch of the evolutionary tree, and their rapid appearance, diversification and disappearance from the world's oceans is an intriguing paleontological puzzle.

 ## THE CAMBRIAN FOSSILS

The very first organisms to have mineralized body parts, and therefore to leave traditional fossils, appear in the Cambrian. The earliest group, called the Tommotian fauna, after a classic locality in Russia, is the beginning of the Cambrian explosion, and seems to appear in the rock record full-blown, with considerable diversity and no very obvious precursors. But even these fossils are somewhat enigmatic. For lack of a better understanding of their true nature, paleontologists just call them "small shelly fossils." It's not clear whether many of these objects, which have a variety of shapes—tiny cones, round, flattened caps, small coiled shells, and many others—are small parts of larger organisms, or major parts of small animals. They appear quite abruptly at the beginning of the Cambrian, quickly reach a peak in abundance and variety, and then decline rapidly, to be supplanted in importance by other animals. But they did attain essentially worldwide distribution, and are found in rocks of earliest Cambrian age in many localities throughout the globe.

Even geologists need occasionally to pause and reflect on what is meant by words like *rapid* and *quickly,* as they are used in the previous paragraph. We draw graphs that plot the number of different forms of the small shelly fossils against time, and they show what does indeed appear to be a sudden appearance and increase, and an equally rapid decrease, early in the Cambrian period. The rise and fall form a small, sharp blip on the graph. But the small blip is at least 10 *million* years in width. Both the "rapid increase" and the spread of these organisms around the globe probably occurred over millions of years. A million years is a long time; a few million years ago the modern Ice Age in the Northern Hemisphere was just beginning (it is still in progress; more on this later) and our species,

Homo sapiens, did not yet exist. Ten thousand years ago, only ⅟₁₀₀ of a million years, great tusked mammoths roamed North America. It takes constant effort to keep geologic time in perspective.

So the spread around the globe of the animals that have left us the small shelly fossils is perhaps not as remarkable as it might at first appear when the true magnitude of the timescale involved is realized. The globe was also a very different place at the beginning of the Cambrian. Although reconstructing the world's geography at such a distant time in the past is difficult, there are enough clues to be reasonably confident about the general outlines. Many of the presently separate continents—Africa, India, Australia, South America—were joined in a single landmass, which may have facilitated the spread of the small, sea-dwelling creatures in the shallow waters along continental margins. However, North America and what is now Siberia apparently were quite separate. For perspective, a view of the world near the beginning of the Cambrian period is shown in Figure 7.1.

The small shelly fossils of the Tommotian fauna are just the first pulse of the Cambrian explosion. Most of the major groups

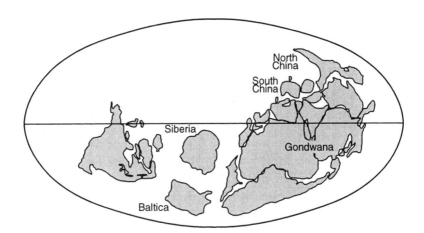

FIGURE 7.1 *A map of the world showing the approximate positions of the continents at the beginning of the Cambrian period. Many of today's continents were joined in the giant landmass of Gondwanaland, although North America was not. Modified after Figure 1 of W. S. McKerrow et al. in* Journal of the Geological Society of London, *vol. 149, page 600. Geological Society of London, 1992.*

of invertebrate animals appear shortly thereafter. These include things like sponges, and the characteristic Cambrian fossil animals, the trilobites. Although now extinct, the trilobites were abundant in Cambrian seas. They were mostly fairly small, and nearly all varieties crawled about on the seafloor, commonly leaving traces of their progress in the soft mud. These too occur as fossils. Trilobites had hard, calcified external skeletons, which undoubtedly helped to protect the soft parts of their bodies from predators, and they are well preserved in many Cambrian sedimentary rocks. They are so common that small specimens can be bought quite cheaply at shops specializing in minerals, rocks, and other items from nature. Figure 7.2 is a sketch showing two typical Cambrian trilobites. For many years we had small trilobites with glued-on magnets for posting newspaper cartoons, kids' drawings, and other artifacts on the refrigerator door. But it seemed an ignominious end for a creature of the Cambrian seas.

Why did animals suddenly develop mineralized skeletons and hard carapaces in the Cambrian? And why did they do so in such

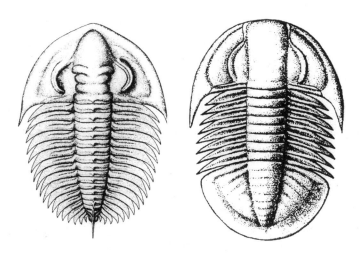

FIGURE 7.2   *Two common trilobites from the Cambrian period. The individuals shown here, sketched from actual fossils, were about 20 (left) and 5 (right) centimeters in length. From Figures 137-1a and 164-2 in* Treatise on Invertebrate Paleontology, *Part O, edited by R. C. Moore. Geological Society of America and University of Kansas Press, 1959. Used with permission.*

bewildering variety? Even those who specialize in research on these questions have no unequivocal answers. Certainly animals with hard armor would have had a better chance of surviving in the face of predators. And skeletons, as well as rigid outer coverings, may also have aided locomotion. The minerals precipitated by the fossil organisms to make their hard parts are calcium carbonate and calcium phosphate. It has been suggested that there was a change in the chemical composition of seawater near the Proterozoic-Cambrian boundary, making it easier for these minerals to form. However, it is hard to conceive of circumstances that could lead to such changes but had not already occurred numerous times during the preceding several billion years of earth history. It is likely that there was no single cause for the Cambrian explosion, and that it occurred when it did through a confluence of many factors.

THE BURGESS SHALE

An interesting aspect of the Cambrian story that bears on this question is that the rapid diversification of animals was not restricted to those with hard parts. If changed ocean water composition was important in the development of skeletons and shells, it was only part of the reason for the amazing diversification of life. Although the traditional view of Cambrian evolution is based on mineralized fossils, in recent years much attention has been paid to the much rarer fossils of soft-bodied creatures. These have been preserved in a number of places where geologic circumstances prevented their rapid degradation. Perhaps the most famous occurrence is the Burgess Shale, in the Rocky Mountains of southern British Columbia. From a single small quarry in this shale have come tens of thousands of fossils exhibiting a stunning range of shapes and body forms, and their study has provided paleontologists with priceless information about the workings of biological evolution. Two of these bizarre creatures are shown in Figure 7.3.

The fascinating tale of the discovery and study of the fossils in the Burgess Shale is told with great skill in the book that provided the title for Chapter 3: *Wonderful Life,* by S. J. Gould. Gould, as already mentioned, is a Harvard paleontologist, and writes with enthusiasm about a subject evidently close to his heart. For anyone

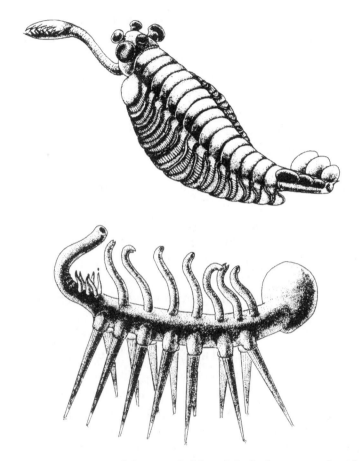

FIGURE 7.3 *Two of the remarkable soft-bodied creatures found fossilized in the Burgess Shale.* Opabinia *(upper diagram) had a strange projecting nozzle and five bulbous eyes, while* Hallucigenia *(lower diagram) presumably "stood" on the seafloor supported on its seven pairs of "legs." These illustrations are reproductions of original drawings by Marianne Collins from* Wonderful Life: The Burgess Shale and the Nature of History *by Stephen Jay Gould. Reprinted with the permission of W. W. Norton & Company, Inc. Copyright © 1989 by Stephen Jay Gould.*

with a yen to learn more about the diverse soft-bodied animals of the Cambrian, and their implications for evolution, his book is highly recommended. Much of what follows draws on it.

Geologists give names to rock formations that have constant or near-constant physical properties and appearance, and can be traced and mapped conveniently over a substantial area. The Burgess Shale is actually just a small, informally named unit within such a formation (the Stephen Formation) in the British Columbia Rockies. The sedimentary bed in which the original discovery of the Burgess fossils was made is only about two and a half meters thick, and it is claimed that this small pod of rock contains fossil evidence of a greater diversity of body plans than exists in today's oceans! Like all shales, it is composed mainly of compacted and hardened fine-grained clay and mud. It was originally laid down in the seas along the western margin of the North American continent near the middle of the Cambrian period.

Because it is well exposed on largely untreed mountain slopes, the original geologic setting of the Burgess Shale has been worked out in some detail by mapping its extent and relationships with surrounding rock types. The animals now fossilized within it apparently lived in relatively shallow coastal waters on mud banks facing the open ocean. These banks abutted tall, calcium carbonate reefs made by algae, some as high as 200 meters, the Cambrian equivalent of coral reefs, which had not yet evolved. But this setting poses a problem for the Burgess fossils: Had the soft-bodied creatures simply died on the mud banks, they would have been devoured by scavengers or, if not, they would rapidly have decomposed. Furthermore, based on their morphology, many of the organisms must have crawled about on the seafloor or burrowed into it, yet no trace fossils of these activities accompany the animals.

The most satisfactory solution to these puzzles seems to be that the Burgess animals (and plants, for they also appear among the fossils) were caught up in small mud avalanches, swept off their aerated and sunlit banks, and unceremoniously dumped into deeper waters where they were buried alive and preserved in an oxygen-poor environment. This scenario would also explain why fossils are rare in much of the rest of the formation of which the Burgess Shale is a part, and why, when they do occur, they tend to appear in great numbers in restricted sedimentary beds.

Assemblages of fossils of soft-bodied Cambrian organisms like those of the Burgess Shale are not abundant, but they have now been found in many other places around the globe. Their scarcity is probably not due to the fact that these animals were rare, but rather because they were not usually preserved.

By the time of the Burgess Shale deposit in the middle Cambrian, some organisms had developed hard parts. These appear as fossils alongside their soft-bodied contemporaries, and because they constitute an assemblage typical of those found in numerous other localities, it is inferred that the soft-bodied animals in the Burgess Shale were also typical of their time. One can say with some confidence that both soft-bodied animals and those with hard parts took part in the Cambrian explosion.

As related by Gould in his book, the story of the Burgess Shale revolves around two men: C. D. Walcott, a geologist and very influential American scientist of his day, who discovered the soft-bodied fossils of the Burgess Shale in 1909, and H. Whittington, a British paleontologist who began a restudy of the location and its fossils in the late 1960s. Walcott was an acknowledged expert on Cambrian fossils. At the time of his discovery of the Burgess Shale, he was also head of the Smithsonian Institution in Washington, D.C. Summer fieldwork in the breathtakingly beautiful Rockies, often with his entire family, was a welcome respite from his administrative responsibilities.

Walcott came upon the Burgess fossils near the end of his 1909 field season, and immediately recognized their importance. His field notebooks show that he concentrated on collecting and describing as many examples as possible over the short time that remained to him that season. For the next four years he returned to the site every summer, and visited it again in 1917. In all, he collected and brought back to the Smithsonian some 80,000 specimens!

Many of the animals that Walcott described and sketched in his notebooks were unique, never before encountered. They showed an incredible diversity of forms. Yet Walcott was a busy man; when he returned to Washington at the end of each field season he was burdened with administrative duties and endless assignments on national committees and commissions. As a result, one can surmise, he was not able to spend a great deal of time thinking about the implications of his Burgess Shale collection. He published descriptions of the fossils, but no detailed, analytical works. In spite of what appear even from Walcott's descriptions to be bizarre body forms, he placed them all within the context of well-known younger fossil and modern animal groups. Gould uses the term "shoehorn" for this approach, and it is an apt characterization: Like

Cinderella's stepsisters, but without the evil intent, Walcott tried to cram into the glass slipper something that didn't really fit. Remarkably, his characterizations went essentially unchallenged for more than half a century, until the work of Whittington and his colleagues.

Shale is rock that characteristically breaks into flat, platy pieces. In places like the Rockies, steep talus slopes of shale provide a wonderful diversion for geologists at the end of the day: With a running start, it's often possible to slide downhill for hundreds of meters riding a mini-avalanche of shale pieces. The abundant flat faces also provide good fossil hunting, and on them the Burgess Shale fossils appear to be flattened, two-dimensional representations of the original animals. This is how Walcott viewed them. It was left to Whittington and his talented coworkers in the Burgess project to reveal the three-dimensional aspects of these creatures, and in doing so to recognize that in fact they could not be shoehorned into extant groups as proposed by Walcott.

Whittington worked on the Burgess material at Cambridge University in the United Kingdom. Together with two graduate students, Conway Morris and Derek Briggs, who are now acknowledged experts in their own right, he turned the conventional wisdom about Cambrian evolution on its ear. These three scholars showed that instead of being an orderly progression from primitive to more advanced forms, with ever increasing numbers of different, specialized groups, Cambrian evolution appears to have been chaotic, truly a game of chance, rife with experimentation that often failed. A great number of the fossils they described and studied seemed to have no recognizable successors among later animals.

Although Whittington and his colleagues collected their own Burgess Shale fossils, Walcott's collection was much bigger and more comprehensive, and many of their discoveries resulted from restudy of the Smithsonian samples. Their advances also depended on the approach they used. With microtools such as dental drills, Whittington and his coworkers painstakingly *dissected* (!) the fossilized animals, some of them only centimeters in size, following appendages layer by layer through the solid rock to determine (for example) the number of joints on a limb, and on occasion even excavating recognizable small organisms from the guts of larger ones, the remains of a Cambrian meal. These very detailed examinations allowed Whittington's group to document the uniqueness of many of the Burgess animals.

No one doubts the amazing diversity found among the Burgess Shale fossils, nor that this diversity arose quickly in a geologic sense. There are, however, differences of opinion about the implications of these observations that are not yet resolved. Some view the plethora of body forms as a first-order paleontological mystery, because many of the animals have neither obvious precursors nor successors in the fossil record. Others are more sanguine about this problem, and suspect it is an artifact of the incompleteness of the record. After all, there are many gaps even in the fossil record of animals with skeletons and hard body parts; the potential for preservation of most of the soft-bodied Burgess Shale creatures is, by comparison, minuscule. Furthermore, near the beginning of the evolution of multicelled animals, high diversity and rapid diversification might well be expected as organisms adapted to the many available ecological niches and living arrangements. Frequent and rapid extinctions would also be expected when some designs didn't work, and still-surviving groups would then have expanded to fill the voids. Perhaps there is more continuity than is apparent, if only the material existed that would permit paleontologists to follow the rapidly changing evolutionary pathways. Currently there is great interest in the Burgess Shale–type fossil assemblages that have been discovered in various places around the world. Perhaps before too long, our view of the Cambrian explosion will be in even sharper focus.

I close this chapter with the fascinating question posed by S. J. Gould: What if the tape were rerun? Regardless of differences in interpretation of the Cambrian fossil record, most experts in this field would probably agree that the result—today's animal world—would be different, quite possibly radically different. Rapid diversification in the Cambrian (and also at some later times), coupled with apparently random extinctions, left groups of survivors that made it more by chance than by any kind of predetermination. Certainly factors such as adaptability and tolerance to a wide range of environmental conditions must have helped, but if the process were restarted it is unlikely that the same groups would persist. The high probability that *Homo sapiens* would never have arisen, a profoundly unsettling concept to some, seems inevitable.

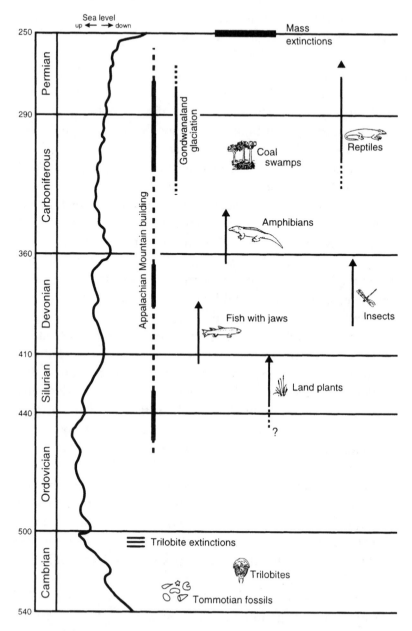

Events of the Paleozoic era, including sea level changes. Time is in millions of years before the present.

8

FISH, FORESTS, AND GONDWANALAND: THE PALEOZOIC ERA

THE CAMBRIAN EXPLOSION ushered in the Paleozoic era, literally the "interval of ancient life." Lasting roughly 300 million years, it saw life evolve from the primitive creatures fossilized in the Burgess Shale to fishes, insects, reptiles, and eventually even the immediate forerunners of mammals. The continents, as well as the oceans, became suitable habitats for living things. Toward the end of the Paleozoic era, with a warm, benevolent climate, widespread lush, swampy forests developed—the source of many of our coal deposits. For much of the Paleozoic most of the modern southern continents—Africa, South America, Australia, India, and Antarctica—were joined in a gigantic landmass, Gondwanaland. Near the end of the era, collisions between Gondwanaland and the remaining continents formed an even larger landmass that geologists have named Pangea, comprising virtually all of today's continents. Pangea stretched from pole to pole, and the collisions involved in its formation threw up great mountain ranges in what are now eastern North America, Scotland, Asia, eastern Australia, and other areas as well. Marking the end of the era was perhaps the greatest mass extinction of all time. No living things completely escaped its influence. Plants and animals, ocean-dwelling and land-dwelling, all were affected. Species and families simply disappeared from the fossil record. Within the families that did survive, there were often only a small number of remaining species. The cause of this global disaster is still unknown.

Paleozoic rocks abound on the continents. Compared to earlier times, there is a remarkably complete record of events during this part of earth history. Partly this is due to the fact that sea level was very high during much of the Paleozoic. Continental interiors were often flooded with shallow seas from which layer upon layer of sediment were deposited. Plants and animals were entombed, preserving evidence about the climate and the environment of their deposition. In North America, for example, much of the interior was intermittently under water during the Paleozoic. These submarine intervals are reflected in the Paleozoic sediments that still blanket much of the continent. Their preservation has been aided by the fact that the nucleus of North America, the foundation upon which the sediments were deposited, is an old, stable shield of Archean and Proterozoic rocks, geologically quiescent and low-lying. Such regions are relatively unaffected by erosion.

The good preservation of Paleozoic and younger rocks, and the myriad of clues they contain about life, climate, and tectonics, make the task of relating the most recent half billion years or so of the earth's history in a concise way quite difficult. There are so many details that there is a very real danger of losing sight of the forest for the trees. For that reason, in this and subsequent chapters only a broad outline of some of the more important events is given, and only a few of these are probed in any detail. You may have noticed from Figure 1.1 that each of the three geologic eras of the past 500 million years—the Paleozoic, Mesozoic, and Cenozoic—is shorter, in actual numbers of years, than the preceding one. This is because, as in most histories, there is progressively more information available for detailing and subdividing geologic history as the present day is approached. For those interested in digging more deeply into this record than can be done here, there is a short bibliography at the end of the book. These sources, and the references they contain, should satisfy most curiosities.

TRILOBITES AND THE CAUSES OF EXTINCTIONS

At the beginning of the Paleozoic, the continents were still almost lifeless. With the exception of algae, which by this time had colonized the land and probably gave the moister regions a verdant tinge, the continents must have been as barren as the surface of the moon. But by the end of the era, forests flourished, insects winged their way through the sky, and reptiles scurried about. Fish abounded in the lakes that populated Pangea. The development of these life forms is reasonably well known, and it is a fascinating story. It is a tale that at times proceeds slowly and uneventfully, but at others is punctuated by very rapid changes.

The preceding chapter described, very briefly, the Cambrian explosion. That event resulted in an oceanic flora and fauna that was without precedent in the earth's history. But it was only a beginning. Very rapidly by the standards of the vast tracts of time that had already passed by, new life forms arose. But also, older ones declined and sometimes quite suddenly became extinct. The precise causes of these rapid changes are not known in detail, but most of them were probably responses to external stimuli. If the world had been completely stable, with an unchanging climate, uniform environments, a constant population density and no continental drift, many of these changes would probably never have occurred. At best they would have taken place much more slowly.

The trilobites, the characteristic Cambrian fossils, provide an instructive illustration of species extinctions and their possible causes. Their history is particularly well documented in North America, which was near the equator during much of the Cambrian and was periodically flooded by warm, shallow seas. Many species of Cambrian trilobites have been identified, and from the types of sediments in which they are fossilized it has been possible to determine something about their lifestyles. They includeed swimming forms, bottom dwellers, varieties that lived in warm, shallow waters, and those that lived in deeper, cooler regions. Most of the individual trilobite species persist in the fossil record for only a few million years or less, although at any given time there were many different species in existence. The regular disappearance of species that had been present for several million years, and the appearance of new species, are part of the background noise of evolution. However, at three separate times toward the end of the Cambrian period, large numbers of the species that were extant disappeared abruptly, apparently on timescales of a few thousand years, or perhaps even less. One of these trilobite "mass extinctions" marks the end of the Cambrian period. After each of them there was a period referred to by paleontologists as one of adaptive radiation, a rapid proliferation of many new species from single progenitor groups.

What caused these sudden events? Throughout the history of life on earth there have been extinctions, many affecting only a few species, others apparently global and catastrophic. Geologists and paleontologists have tried to uncover the reasons for these extinctions from the record in the rocks. For the most part they have not discovered any unequivocal answers, but there are a number of recurring themes. These include changes in climate, shifting continents, the evolution of predators, and changes in sea level. In the specific case of the trilobites, there are indications that climate change was involved. The extinctions appear to have been most severe for species that lived in the warmest waters. Furthermore, the ancestors of most of the species that evolved rapidly (radiated) after each extinction were trilobites that lived in the cooler, deeper waters along the margins of the continents, leading many geologists to infer that the extinctions occurred due to sudden cooling. The cool-water species were able to withstand the change; those that were adapted to warmer water died out. Although this hypothesis is by no means proven, it is a reasonable deduction from the available evidence. This example from the trilobites is not unique; there are many places in the rock record where the evidence indicates that climate change has had important consequences for the course of evolution.

However, not all changes in the flora and fauna during the Paleozoic resulted from climatic variability, at least not directly. The stromatolites, those structures built up by layer upon layer of algal mats that were so widespread in the Proterozoic, decreased markedly in abundance during the Paleozoic. But in their case there is no evidence that the decline was in any way related to climate change. Instead, it is generally attributed to the appearance of predators that fed on algae, and of burrowing animals that destroyed the algal mats as soon as they formed. The stromatolites alive today grow only in restricted environments where such creatures are rare or absent.

Many extinctions seem to be related to changes in sea level, although there is little evidence that this was the case for the trilobites, or the decline of the stromatolites. However, even relatively small changes in the level of the oceans would drastically alter habitats along coastlines and in shallow inland seas. They therefore have the potential to cause great disruptions among organisms living in such environments. But how do we know about sea level changes hundreds of millions of years ago? Obviously there was no one around to make measurements, but once again the clues are present in the rocks. They tell an interesting story, and the following section—a minor diversion from the details of Paleozoic history—illustrates the kind of reasoning that has been used to ferret out the details of the rising and falling ocean levels.

CHANGES IN THE
LEVEL OF THE SEAS

During much of the Paleozoic, the rock record shows, sea level stood quite high relative to the continents. But there were also substantial fluctuations. Continental flooding was pervasive, but it was intermittent. Some particularly well documented evidence for changing sea level comes from the rocks of western North America.

The key to understanding how sea level changes are recorded in rocks is fairly simple: Most sedimentary rocks were originally deposited in water, and all bodies of water accumulate sediments at their bottoms. Rainfall and erosion ensure that the rocks of the continents will be worn down, and the detritus of that process is carried by rivers to lakes and the ocean, where it settles out in layers of sediment. Furthermore, the coarse material drops out first, close to the shoreline, while the finer grains remain suspended and are carried further out to sea. Thus the type of sediment that is formed depends on the water depth. Even by using just these fairly obvious principles, a great deal can be learned about the sea level changes that occurred in the Paleozoic.

The sedimentary rocks of western North America are particularly instructive because in the early part of the Cambrian, and during the time the Burgess Shale was being deposited, the region now occupied by the Rocky Mountains was at the edge of the continent. Reconstructions of the world's geography show that the position of the continent was actually rotated in the Cambrian compared to today, and what we now know as western North

America actually lay along its northern margin, close to the equator (see Figure 7.1). But that is really only incidental to our story. The important point is that most of the western part of the continent was fairly stable—there were no collisions with other continents, and no widespread volcanism associated with the subduction of an oceanic plate. Sediments were deposited along this quiet continental edge during most of the Paleozoic. If you have ever been to the Grand Canyon, you have seen the results of this process: The flat-lying sedimentary rocks that make up most of the canyon walls range in age from Cambrian to Permian, a time span that encompasses the whole of the Paleozoic era. And if you look closely, you will notice that a variety of rock types are represented, each reflecting a different environment of sediment deposition. In fact, part of the beauty of the canyon results from this variety, because each rock type erodes in a different way, some making ledges and sheer cliffs and others weathering to gentler slopes. What is not so obvious to the casual observer is the fact that there are large gaps in the sediment sequence of the canyon walls, considerable spans of time during the Paleozoic with no rocks at all to represent them. Both these missing pieces and the variation in rock types are at least partly the result of fluctuations in Paleozoic sea level.

It is fairly obvious that sand is a common type of sediment at the edge of a continent. Along most seashores you don't have to travel far to find a sandy beach, however small. Coastlines are active places; ocean waves provide lots of energy to transport and winnow the material delivered to them by the rivers, and very fine-grained sediments such as clay, which is the main mineral in shale, simply don't accumulate there. The relatively coarse-grained sand accumulates near the shore, while fine particles remain in suspension and are carried out to quieter, deeper water, where they slowly settle to the bottom as mud, which, over long time periods, hardens to become shale. Even farther from shore, most of the fine-grained material has already settled from the water column, and the bulk of the accumulating sediment is likely to be the remains of the marine organisms that live near the sea surface and make their skeletons and shells of calcium carbonate. Such deposits eventually become limestone. Thus if you were to trace the sediment types from the shoreline out into deep water, along many coastlines you would find a sequence consisting of sand and sandstone near the coast, mud or shale farther out, and finally sediment composed largely of calcium carbonate. Now imagine what would happen if sea level were to rise. The whole sequence would shift toward the interior of the continent. The sandy sediments of the earlier shoreline would no longer be at the very edge of the sea, but in deeper water. Shale would be deposited on top of these beds. Further rise in sea level would result in even deeper water at the old shoreline, with deposition of calcium carbonate-rich sediments on top of the shale. Thus the shoreline-to-deep-water succession of sandstone, shale, and limestone would, over time, become a *vertical* sequence of the same rock types at the position of the old shoreline. This is exactly what is observed at the Grand Canyon. But can we really use the principle of uniformitarianism, and interpret the vertical sequence of sediments in the Grand Canyon in terms of a gradual change in water depth along the margin of the Paleozoic continent? The answer is an emphatic yes, because throughout western North America the same sequences are observed both horizontally and vertically. Geologic maps drawn for a single point in time show a similar procession of sediment types to that found today along coastlines such as the Gulf Coast of the United States: sandstone, shale, and limestone in an east-to-west sequence. Maps made for earlier or later times show the same sequence, but shifted geographically, reflecting the fact that as sea level rose and fell, the shoreline migrated east and west. These features are illustrated in sketch form in Figure 8.1.

But even if these observations are accepted as a record of rising and falling sea level, there is still the question of what actually changed: Was it the elevation of the continent or the absolute level of the sea? After all, the sediments only record relative changes, and we know that the continents undergo vertical movements—rocks high in the Alps and the Rockies contain fossils deposited in the ocean, for example, and we know that the oceans were never *that* deep. However, the case for true sea level change can be made convincingly if evidence of the sort just described for western North America is found in rocks of the same age from geographically widespread regions. Geologists

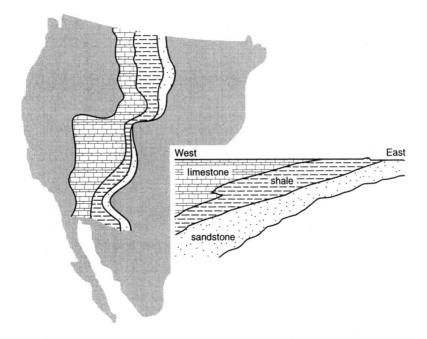

FIGURE 8.1   *The distribution of sandstone, shale, and limestone in the western United States shown in the map view (left) reflects the water depth at a particular instant of geologic time early in the Paleozoic era, with sandstone marking the continental edge and deeper water to the west. In cross section (shown here with an exaggerated vertical scale) age increases downward, and a horizontal surface marks the distribution of rock types at any particular time. Thus it can be seen that the shoreline gradually moved from west to east. Modified after Figures 19-3 and 19-4 in* Earth, Time and Life, *2nd edition, by C. W. Barnes. John Wiley & Sons, 1988. Used with permission.*

have mapped the occurrence of various sediment types almost everywhere on earth in considerable detail, and through syntheses of such data there is now a fairly good understanding of the magnitude and timing of global sea level changes throughout the entire Phanerozoic. A summary of this information for the Paleozoic era is shown on the timescale facing the first page of this chapter. It is obvious from this graph that sea level was high during much of the era.

If the rock record indicates that there have been large changes in sea level, the obvious question is, Why? As far as we know there are really only two possibilities: There must have been changes in either the volume of water in the oceans itself, or in the volume of other things that displace the water, such as continents, or islands, or ocean ridges. For example, we know that glacial periods are characterized by sea level lowering, because large amounts of the earth's surface water are tied up in ice sheets on the continents. It is estimated that at the height of the last glacial advance, roughly 20,000 years ago, sea level was well over 100 meters lower than it is today. And although much of that ice is gone, there is still a considerable amount of water frozen in the ice caps. If all of it were to melt, sea level would rise by about another 65 meters. That may not sound like much, but a large fraction of the earth's population lives close to sea level. Mexico City would be spared, but much of Los Angeles, New York, Tokyo, and Berlin (to cite just a few examples) would be inundated.

Although glacial episodes have a major effect on sea level, most of the fluctuations that are recorded in Paleozoic rocks don't occur at times for which there is independent evidence for global ice ages. Most likely they were caused by variations in the volume of the oceanic ridges. As described in Chapter 5, hot magma wells up along these features, creating new seafloor. Because new ocean crust is hot and buoyant, the ridges are regions of elevated topography. In fact, their depth is only about half that of the older, colder parts of the seafloor. When the average rate of seafloor spreading increases—either because new ridge segments form, or old ones spread more rapidly—the volume of the ridge system as a whole also increases. This is like putting a brick in a pail of water—the water level rises accordingly. It is probable that the generally high sea levels of the Paleozoic were due to a ridge system that was considerably more voluminous than today's.

THE GREAT CRASH: BUILDING THE APPALACHIANS

The sediments that were deposited in western North America during the Paleozoic, faithfully and more-or-less continuously recording the rise and fall of sea level, were accumulating along what geologists refer to as a passive margin: a continental edge that is completely within a major plate, free of colliding continents, subduction, or volcanism. Such is the state of the east coast of North America today. Late in the Paleozoic era there is evidence in the west for collisions with small fragments of volcanic crust, probably material like the island arcs that characterize the western Pacific today, but there is no indication of a major continent-continent collision. However, the situation on the opposite side of the continent was very different. In eastern North America there is abundant evidence of volcanism, collisions, and mountain building throughout the Paleozoic, part of a process that joined together all of the major continents then in existence to form the supercontinent of Pangea.

The record of this activity is the Appalachian mountain belt. This geologic province stretches from Newfoundland in the north to Alabama in the south, and that is only the part exposed above-ground (see Figure 4.3). Much of the original range is now buried. As you might expect for such a large-scale feature, the Appalachians are quite varied along their length, the result of differences in the details of the geologic history of their various parts. But details aside, the big-picture view of the Appalachians is that they record the closing up of an ancient ocean basin and the suturing together of major continental pieces—North America, Europe, and Africa—into a single landmass. Although these events took place hundreds of millions of years ago, close comparison of the surviving rocks of the Appalachians with those of much younger mountain ranges such as the Alps, which were also formed by collisions of continents, shows many similarities.

The Appalachians today are not a great mountain range with jagged, snowcapped peaks, but instead, for the most part, a series of pleasant, subdued hills and valleys. Included in the Appalachian geologic province are the beautiful Blue Ridge mountains of Virginia, the Great Smoky Mountains of North Carolina and the Green Mountains of Vermont. The present-day topography actually has little to do with the original mountains, which had already been worn down by erosion by the middle of the Mesozoic era. In the southern part of the province, at least half the width of the eroded range is now covered by the low-lying sediments of a coastal plain. Today's mountains result from fairly gentle uplift of the ancient, folded rocks in the geologically quite recent past, followed by differential erosion of the various rock types to form the typical valley and ridge structure of much of the range.

Although the Appalachian Mountains are geologically complex, they have been studied in great detail by American and Canadian geologists for well over a century, and their physical features are well known. However, theories for their formation and evolution that were developed before the advent of plate tectonics in the 1960s are unconvincing. In particular, these earlier scenarios lacked mechanisms that could explain the volcanism, faulting, and intense metamorphism recorded in the surviving Appalachian rocks. With the realization that the continents and oceans are not immutable features of the earth, more plausible hypotheses began to emerge. In its simplest outlines, the currently generally accepted view of how the Appalachians formed runs approximately as described below. However, you should recognize that in reality the history is very much more complex than this brief summary will allow, and that the sequence of events varied greatly from place to place along the vast length of the forming Appalachian chain.

As sea level rose during the Cambrian period and into the Ordovician, the coastline in eastern North America gradually moved to the west. Thick sequences of sediments, especially carbonate rocks such as limestone, were deposited along the edge of the continent. During this time the east coast, like the west, was still a passive margin. But near the middle of the Ordovician period, the ocean to the east began to close up because of seafloor subduction (see Figure 8.2). Before long, the first of three major mountain-building episodes that have been identified in the history of the Appalachians occurred when the last of the intervening seafloor was subducted. The North American continent collided with a variety of crustal fragments, crumpling the great platform of carbonate sediments that had been deposited along the margin, and shoving parts of it far westward toward the continental interior. To the east, on the conveyor belt of plate tectonics, was more ocean, which continued to close as the crust was subducted. Eventually, probably around 380 to 390 million years ago, this process completely swal-

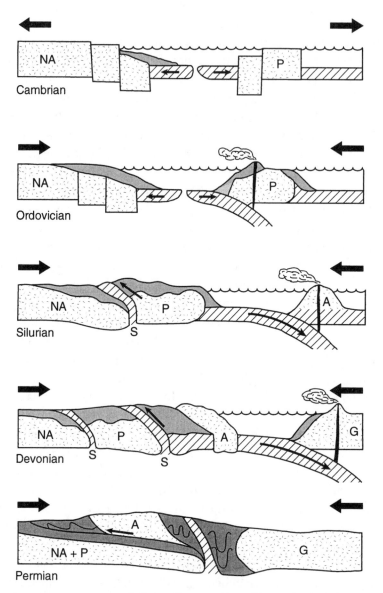

FIGURE 8.2   *A simplified reconstruction of how some of the major mountain-building episodes in the history of the southern Appalachians may have occurred. Letters on the various crustal blocks refer to North America (NA), Gondwanaland (G), and various island arcs or micro-continents that collided with North America (P and A). Recognized suture zones (S) separate these different blocks. After Figure 13.28 in* Evolution of the Earth, *5th edition, by R. H. Dott, Jr., and D. R. Prothero. McGraw-Hill, Inc., 1994. Used with permission.*

lowed up the ocean and what is now Scandinavia, in addition to parts of Great Britain, collided with North America. This second episode of Appalachian mountain building is recognized in Europe as well as in America; the rock types, fossils, and geologic structures from this time are very similar on both sides of the Atlantic. Some 70 or 80 million years later, yet another large continent—northwestern Africa (and probably South America as well)—collided with southern North America, initiating the last of the three major Appalachian mountain-building periods. This collision was probably also responsible for the Ouachita Mountains of Oklahoma and Arkansas, which are essentially a continuation of the Appalachians around the southern margin of North America. In very simplified terms, Figure 8.2 shows a series of snapshots illustrating how these collisions may have proceeded. Note that this sketch refers specifically to events in the southern Appalachians, and shows collision with an island arc (A), rather than northern Europe, as the second stage of Appalachian mountain building.

The final pulse of mountain building in the Appalachians sutured together the giant southern continent known as Gondwanaland, of which Africa was a part, and the northern, America-Europe landmass, one of the last steps in the assembly of the pole-to-pole megacontinent of Pangea (see Figure 8.4). Much later, as we shall see, this giant continent split apart again to form the present-day Atlantic Ocean.

The processes that formed the Appalachians are typical of those that have produced younger mountain belts, such as the Alps. And although the details are fuzzy, many much older geologic provinces, such as the Grenville, which was discussed in Chapter 4, are probably the result of very similar processes. In fact, the Grenville and Appalachian provinces are parallel, adjacent belts (Figure 4.3), both produced by collisions along the eastern edge of North America, and both adding new crust to the margin of the continent. The age structure of the North American continent, with approximately concentric bands of progressively younger crust surrounding ancient continental nuclei, has led many geologists to embrace the view that the growth of continents is a process of gradual accretion at the margins.

The Appalachians were not the only mountains formed during the Paleozoic. The Urals, in central Russia, are also the result of continent-continent collision near the end of the era, yet another step in the assembly of the giant landmass of Pangea. Obviously, the Urals do not now parallel a continental margin as do the Appalachians; the suture held, and the continent has not yet split apart again. Much of eastern Australia was added to the rest of that continent during the Paleozoic as well, through a series of mountain-building events that also affected the Antarctic. Here the process was not one of collision between two major continents, but rather the accretion of island arcs and marginal sediments, much like the first of the three Appalachian mountain-building events.

 ## LIFE IN THE PALEOZOIC

As the processes of plate tectonics arranged and rearranged the continents over the earth's surface during the Paleozoic, culminating in the assembly of Pangea, the evolution of life-forms continued apace, almost certainly strongly influenced by the changing dispositions of land and sea. The repeated extinctions and radiations of the trilobites have parallels in the fossil records of many other groups of organisms from the Paleozoic.

Vertebrates—animals with backbones, like us—are not found among the fossils of the Burgess Shale or its equivalents in other parts of the world. However, they evolved quite early. Their first representatives in the geological record are fishes. Fragmentary fossils that are believed to be parts of fish are found in sediments from near the end of the Cambrian period, and also from the Ordovician. These early fish were apparently strongly armored creatures; many of the fossils that have been found are pieces of bony exterior plates. They seem to have been bottom dwellers that fed by filter feeding, not predators like modern fish, and they lacked biting jaws. A few descendants of these jawless fish still survive today, an example being the lamprey "eel."

Although fossils of the earliest fish are found in ocean sediments, much of the record from the Silurian period onward, starting some 440 million years ago, comes from freshwater rather than marine deposits. Indeed, there is considerable debate about whether the vertebrates actually evolved in freshwater or in the oceans. Unfortunately, the sedimentary record from lakes and rivers is even less complete than that from marine environments, and there is no convincing evidence to settle this question.

By the end of the Silurian period, still more than 400 million years ago, a new group of fish had appeared both in freshwater and in the oceans. These had scales, and abundant sharp, spiny fins. They also had jaws, and it is clear that they must have been effective predators. During the Devonian period the jawed fishes flourished and diversified, and they constituted the main ingredients of a complex food chain in which smaller species were preyed upon by larger species, which in turn were eaten by their even larger brethren. One bizarre Devonian fish, heavily armored with large, bony plates on its head and the frontal parts of its body, reached a size of about 10 meters—truly a terrifying denizen of the deep.

The development of jaws was an important step in the evolution of fishes and indeed for all vertebrates. It is also an interesting example of an oft-encountered feature of evolution, the alteration of some preexisting body part or structure to perform a new task. Most paleontologists believe that the jaws of fishes evolved from the cartilaginous gill supports that existed in the heads of the jawless fishes. They were in the right place anatomically, and with a minimum of change could function as simple jaws, as shown in Figure 8.3. The earliest teeth were probably modified scales. The story of evolution is replete with such truly fascinating details, and one cannot but return again to S. J. Gould's question: If the tape were replayed, would the story remain the same? Would jaws have developed in the same way? Would there even have been fish as we know them in the Paleozoic?

One of the varieties of fish that developed during the Devonian was the ancestor of the terrestrial vertebrates. This group includes the lungfishes, a few examples of which survive today in arid environments in Australia, Africa, and South America. The lungfishes are able to obtain oxygen both directly from the water, via gills as do other fishes, and also by gulping air into rudimentary lungs when the ponds they inhabit stagnate or dry up. The precursors of terrestrial vertebrates had similar capabilities. Ironically, the development of land-dwelling creatures came about through adapta-

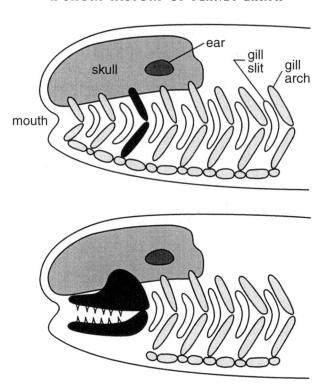

FIGURE 8.3   *It is believed that the jaws of fishes (dark structure in lower diagram), and therefore of vertebrates in general, evolved from gill supports, as illustrated. Teeth, added later, were probably modified scales. Reproduced from Figure 13-12 in* Earth and Life Through Time, *2nd Edition, by S. M. Stanley. W. H. Freeman and Company, 1989. Used with permission.*

tions made by fish to carry them through periods of drought, so that they could continue their lives in water!

The first step in the development of vertebrates that lived entirely on land was the evolution of amphibians. Modern examples such as frogs and toads begin their existence in water, but crawl onto the land when they mature and usually spend most of their adult lives there. Presumably the life cycles of the early amphibians were quite similar. They first appear in the fossil record in the Devonian period, and the details of the body structures of some of the early fossilized amphibians are so similar to those of fishes from the same time period that there is no doubt about their close relationship. They appear to be direct descendants of the group of fishes that include the air-breathing lungfish.

The transition from fish to mostly land-dwelling amphibians apparently took at least fifteen million years, and perhaps longer. As paleontologists have gathered more information about this evolutionary step, the simple idea that a single lineage of fishes adapted to shallow water environments and then crawled out onto the land has given way to a much more complex scenario. As in the Cambrian explosion and at many other junctures in the evolution of life, there were apparently numerous simultaneous and parallel evolutionary branches along which the amphibians developed. In spite of similarities in body structures and other features among these lineages, only a few survived.

However, the successful amphibians had the land to themselves, and diversified rapidly. There were some problems to overcome, of course: Their predecessors were constantly bathed in water, and land was a completely foreign environment. They had to develop ways to avoid desiccation, and to move around without swimming. Furthermore, their skeletal systems had to become much stronger to support their full body weight in air, a much less dense fluid than water and therefore less supportive. And they had to have lung systems that would permit them to spend most of their lives out of water. Nevertheless, amphibians prospered, some of them becoming quite large. They included both meat eaters and herbivores. But by the end of the Paleozoic era they had been replaced in importance by the reptiles, setting the stage for the rise of the dinosaurs. A key evolutionary event that favored the reptiles was the development of an egg like those of reptiles and birds today, with a tough outer covering and an internal supply of nourishment—an egg that didn't have to be laid in water. In effect, the reptilian egg was its own portable pond, bathing the growing embryo in a friendly fluid during the critical early stages of its development. It permitted the parent reptiles to enjoy a much less restricted lifestyle.

The first reptiles appear in the fossil record about 330 million years ago, in the Carboniferous period. Long before this, and even before the amphibians had evolved, plants had already invaded the land. Like their animal counterparts, they proliferated rapidly to fill the new environment. In fact, the Carboniferous period is named for its abundant, carbon-rich coal deposits, the altered remnants of huge masses of plant material from ancient tropical forests.

The earliest plant fossils appear in the Silurian period. Like the amphibians and reptiles, plants faced considerable problems in colonizing the land, problems that were in fact very similar to those of the animal kingdom. The earliest plants bore spores, like modern ferns, and required moist conditions to reproduce. They were probably restricted to near-shore environments, living partly within, or at least close to, bodies of water. Spore-bearing plants proliferated during the Paleozoic. Most of the Carboniferous coal deposits of the Northern Hemisphere for example, are largely the remains of huge, spore-bearing trees that grew in swamps. A peculiar feature of many of these deposits is that they occur in repeated cycles of coal beds alternating with marine sediments. Apparently the low-lying coal swamps were periodically inundated with seawater, another indication of fluctuating Paleozoic sea level. There is evidence of high-latitude glaciation in the late Paleozoic, and many geologists believe that the geographically widespread occurrences of these alternating coal and marine sediment deposits are the result of waxing and waning of the polar ice caps, with consequent changes in sea level.

Like the animals, plants faced the necessity of stronger structures to support their own weight in air as they moved out of their watery environments onto the continents. This eventually led to the development of thick stems and wood. Most important was the evolution of a system to transport water and food through these stems—the so-called vascular system. And in a development that in some ways parallels the emergence of the reptilian egg, the event that permitted plants to spread widely and rapidly across the continents was the development of seeds. Seeds appeared in the Devonian period, and they allowed plants to reproduce without the necessity of wet conditions. Very soon large seed plants with thick, woody stems and extensive root systems—trees—evolved, and the appearance of the land surface was transformed. Soil, as we know it, with a high content of organic material from decaying vegetation, appeared for the first time in the history of our planet as plants colonized the continents. Although the plants and animals that were present at the end of the Paleozoic era were very different from those of today, the earth would have been a much more familiar place at the end of the era than at the beginning.

Perhaps the most familiar life-forms present at the end of the Paleozoic would have been insects. They appear in the fossil record in the Devonian period, not too long after the early land plants developed, and before the amphibians had begun to populate the continents. The earliest insects were wingless, but by the end of the Paleozoic there were dragonflies, grasshoppers and—believe it or not—cockroaches, which, like the coal swamps, flourished during the Carboniferous period. The habits and habitats of insects were widely diversified even in the Paleozoic, and their evolution must have been closely intertwined with the development of terrestrial plants and animals. Today the insects are the most numerous creatures on earth.

An interesting aspect of the movement of life onto the land has to do with something that is much in the news today—the ozone layer. Ozone is a molecule made up of three oxygen atoms. Most oxygen in the atmosphere is O_2, but energetic radiation from the sun breaks down some of this oxygen into individual atoms in the upper parts of the atmosphere. Ozone is formed when these atoms combine with remaining O_2 molecules to make O_3. The importance of the upper atmosphere ozone layer for life on earth is that the O_3 molecule absorbs short-wavelength (ultraviolet) radiation from the sun. Without this shield, most life on land would be severely affected, if not completely exterminated, by intense ultraviolet radiation. Life in the oceans is much less sensitive, because even a relatively small covering of water provides an efficient shield, blocking the damaging radiation.

In Chapter 4, evidence from the rock record was described that indicates an increase in atmospheric oxygen around two billion years ago. Nevertheless, most geologists believe that even by the early Paleozoic, the oxygen level had reached only a small fraction of its present value. As it turns out, the maximum ozone production in the upper atmosphere occurs when the oxygen content is approximately 10 percent of today's level. At this concentration of atmospheric oxygen the ozone layer provides the most effective

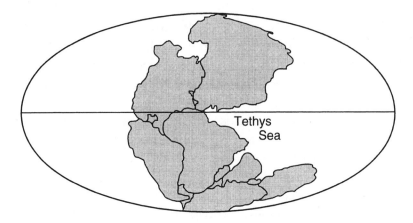

FIGURE 8.4 *At the end of the Paleozoic, all of the major continents were joined in the giant landmass known as Pangea, which stretched from pole to pole. Modified after Figure 20-17(a) in* **Earth,** *4th edition, by F. Press and R. Siever. W. H. Freeman and Company, 1986.*

protection against lethal ultraviolet radiation. To the best of our knowledge, this is approximately where it stood in the Silurian period when the earliest plants appeared on the land. Could this be a coincidence? Probably not.

It is obvious from even the few tidbits of Paleozoic history that it has been possible to relate in this chapter that the era saw revolutionary changes in the nature of the earth. By its end, life had colonized the continents, which now were almost all joined in a landmass that extended from pole to pole (see Figure 8.4). Major mountain-building episodes had accompanied the assemblage of this gigantic continent, throwing up ranges such as the Appalachians and the Urals. The stage was set for the evolution of mammals, dinosaurs, and birds in the Mesozoic era. But the Paleozoic closed—with apologies to T. S. Eliot—not with a whimper, but with something much more cataclysmic. The boundary between the Paleozoic and the Mesozoic eras is marked by the most widespread extinctions known in the fossil record. It is estimated that between about 80 and 90 percent (!) of all species living in the oceans at the end of the Permian period did not make it into the Mesozoic, and while the record is much less complete for terrestrial plants and animals, it is apparent that they too were severely affected. Although there are many theories, there is no certainty about the causes of this disaster for life on earth. Some of the ideas that have been proposed to explain this and other mass extinctions in the geologic record will be explored further in Chapter 10.

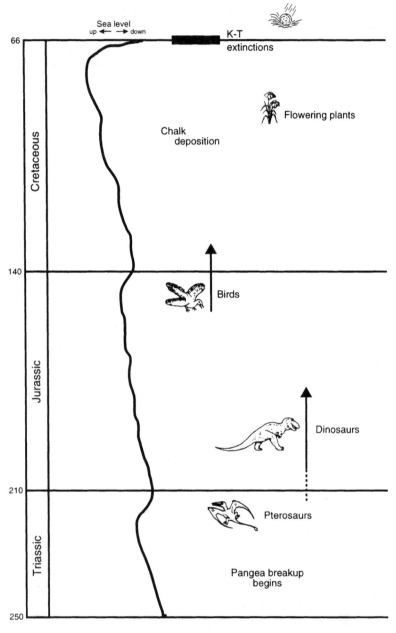

Events of the Mesozoic. Time is in millions of years before present.

9

FROM PANGEA
TO (ALMOST) THE
MODERN WORLD:
THE MESOZOIC ERA

WHEN THE MESOZOIC ERA began some 250 million years ago, most of the current continents were united in the giant landmass Pangea, as discussed in the previous chapter. By its close, 66 million years ago—not so long ago, geologically speaking—the physical world looked much more like today's. There were still some significant differences, of course: India was a great island south of the equator, moving north toward its eventual collision with Asia, and Australia was still attached to the Antarctic continent. But a map of the world 66 million years ago is not completely unrecognizable.

The Mesozoic is sometimes referred to as the age of reptiles. Although a variety of reptiles flourished then, the most characteristic, especially during the Jurassic period, were the dinosaurs, as anyone who has read the book *Jurassic Park,* or has seen the film, can attest. The close of the era, like the close of the Paleozoic, was marked by a major mass extinction. It was not as severe as the one at the end of the Paleozoic, but because it includes the dinosaurs, it has captured the imaginations of scientists and nonscientists alike. Although the dinosaurs were very successful animals, their end was exceedingly abrupt, more so than most other extinctions in the geologic record. As will be told in the next chapter, the end-of-the-Mesozoic mass extinction appears to have been the result of a very sudden global catastrophe. Were it not for this unlucky event, the dinosaurs might well be with us still.

 ## PANGEA, CLIMATE, AND THE BREAKUP OF A SUPERCONTINENT

There is considerable evidence in the rocks that near the very end of the Paleozoic era, during the Permian period, sea level decreased to quite low levels. Because the continents were joined together in Pangea, there was relatively little newly forming seafloor along ocean ridges. Young ocean ridges bulge upward to quite shallow depths, displacing ocean water onto the continents; in contrast, older seafloor sinks to deeper levels with the opposite effect, and this may be the reason for the low Permian sea level. There is also evidence that at the end of the Paleozoic and into the Mesozoic, the climate was quite dry, particularly in the interior of the Pangea continent. Part of this evidence comes from the flora and fauna preserved as fossils, and part comes from the types of sedimentary rocks that were formed at this time.

What are the clues in the rocks to an arid, warm climate? An important one is an abundance of sandstones, in particular sandstones that are made up of lithified sand dunes. In today's world sand dunes characterize the hot, dry conditions of deserts, and there is no reason to believe that this was different in the past. Sandstones can originate in several different environments, but it is usually fairly easy to distinguish between those that were once sand dunes and those that were deposited along beaches or in rivers. For example, wind does not transport large grains and pebbles efficiently, so the grain size of the particles in sand dunes is small and much more uniform than is the case for beach or river sands. In addition, the bedding structures, which reflect the way in which the deposits were actually laid down, are quite different in the two cases. But although dune deposits are common in the Permian, they are not the only evidence for aridity. There are also abundant evaporites, or salt deposits, formed when bodies of seawater become isolated from the open ocean and simply evaporate away, leaving behind only the salts that had been dissolved in them. Like sand dunes, the evaporite deposits signify warm, dry climatic conditions.

There is some debate among geologists about the significance of the evidence just discussed for understanding the global climate at the end of the Paleozoic. Pangea straddled the equator, and many of the evaporites and dune deposits were formed at low latitudes. Perhaps the climate was not unusually warm, and their presence is just the result of their geographic location. Furthermore, the large continental landmass of Pangea presumably had arid conditions in its interior, with large swings in temperature—hot summers and cold winters—regardless of the global average. So in spite of the evidence, one has to be careful about its interpretation. It is not a simple matter to reconstruct the details of the earth's climate a quarter of a billion years ago with any great accuracy.

But regardless of the details, we know that the continents move slowly and the breakup of Pangea took a long time. The influence of this very large landmass on the climate persisted well into the Mesozoic. Evaporite deposits, common as they are in the Permian, are even more abundant in the Triassic. Those of the Triassic, however, not only record warm, arid climates, they also document the initial stages of the splitting up of Pangea. As the supercontinent slowly rifted, the sea periodically flooded into the developing rift valleys. Either due to changing sea level, or because access to the sea was cut off for other reasons, these flooded rifts sometimes dried up, especially if they were located in warm regions, leaving behind characteristic salt deposits. In more recent times, exactly the same process operated in the Red Sea, which is a still quite young rift between Egypt and Saudi Arabia. During its early stages, this rift too was occasionally flooded with ocean water, which then evaporated. The record of these encroachments of the sea is a series of salt layers that underlie the more normal sediments on the floor of the Red Sea.

The breakup of Pangea was the main geographic event of the Mesozoic era. Although it progressed slowly, it was also reasonably continuous. It began with the splitting apart of Europe and Africa from east to west, proceeded to the gradual opening of the North Atlantic Ocean between North America, Europe, and Africa, and finally led to the separation of South America from Africa to form the South Atlantic. By the end of the process, the physical world had been transformed, and this geographic rearrangement had important consequences for both climate and the course of biological evolution. Ocean circulation, the main mechanism for heat transport from one region to another on the earth's surface, was radically altered by the redistribution of continents. Newly formed ocean basins became barriers to plant and animal life, particularly for land-dwelling forms but also affecting marine species. Because the breakup of Pangea is so important for our modern world, influencing everything from the distribution of animal life to our present climate, it is worth following its progress in some detail. "Snapshots" of world geography at intervals during the Mesozoic are shown in Figure 9.1 as an aid in tracing this process.

At the end of the Paleozoic era part of the globe-encircling ocean poked westward into Pangea in the region that we now know as the Mediterranean. Eventually this incursion extended farther westward, splitting apart Pangea and separating Europe and Africa. The resulting east-west-trending body of water became a sea in its own right, known to geologists as the Tethys Ocean or Tethys Seaway. Its formation had a substantial effect on world climate, because it permitted east-west ocean currents to flow through it, and as rifting proceeded even farther westward, sundering South America from North and Central America, the bodies of water on either margin of Pangea were ultimately linked. The rift basins that formed as the Tethys Ocean opened were located in warm, low latitude regions, and as seawater periodically flooded into them during the early stages of their development, and then evaporated, salt deposits formed. These Triassic evaporites can be found today along the northwestern edge of Africa, and in many parts of Europe.

The evaporite deposits of the Mesozoic are useful for tracing out the way in which Pangea split up. By determining their ages, geologists have been able to follow the progression of rift formation and thereby the chronology of continental breakup. The relationship between the early rifting of Pangea and salt deposits was first clearly delineated in 1975 by Kevin Burke, then of the State University of New York at Albany. A sketch showing the locations of the evaporites identified by Burke as having originated with the rifting of Pangea, and the pathways by which seawater entered them, is shown in Figure 9.2. First you should imagine the present continents of North America, Europe, Africa, and South America nestled together as in Figure 8.4, with no Atlantic or

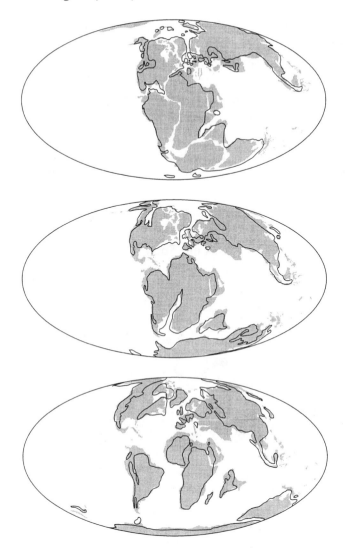

FIGURE 9.1 *The breakup of Pangea was a dominant feature of world geography during the Mesozoic. From top to bottom, these three "snapshots" show the distribution of the continents at approximately 170, 120, and 70 million years ago. To aid in visualizing the process, the present continental outlines are shown in gray shading, while the continental edges as they would have appeared in the Mesozoic are indicated by the heavy dark line. Diagrams modified after maps in* Atlas of Mesozoic and Cenozoic Coastlines, *by A. G. Smith, D. G. Smith, and B. M. Funnell. Cambridge University Press, 1994. Used with permission.*

Mediterranean Oceans. As has already been described, this was the situation at the end of the Paleozoic era. Then breakup began as the Tethys Seaway opened toward the west. Evaporites were deposited in the narrow rifts that preceded the full development of this seaway into the interior of Pangea; their age is early Triassic and they are among the oldest of the Mesozoic salt deposits. The North American equivalents of these deposits, formed as the rifting began to split apart North America from Europe as well as from North Africa, occur along the continental shelf of eastern Canada.

With time, the rifting continued to move west and south, passing through what is now the Gulf of Mexico region of the United States, and eventually breaking South America away from North

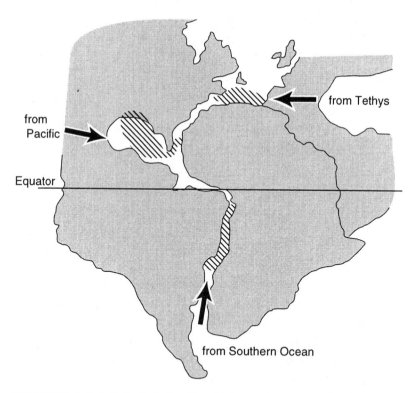

FIGURE 9.2   *Salt deposits (hatched regions in the figure) formed when seawater spilled into the rifts that formed as Pangea began to break up. Modified after Figure 1 of K. Burke in* Geology, *Nov. 1975, p. 614. Geological Society of America.*

America. It was not until well into the Jurassic period that the great salt deposits of the Gulf of Mexico were laid down. It is thought that the water from which these evaporites precipitated entered the rift from the Pacific Ocean side of the continent. The deposits occur both offshore, within the sediments of the Gulf itself, and on land, buried beneath marine sediments in Texas, Louisiana, and Mexico. The evaporites of the Gulf of Mexico are especially well known to geologists, because in some places the salt, due to its low density, has risen in great blobs through the surrounding sedimentary rocks, distorting them and forming traps where large quantities of petroleum have accumulated. These so-called salt domes, which are buried below ground level and detectable only by remote sensing methods, are therefore attractive targets for oil exploration.

Yet another group of evaporites formed as seawater spilled into the early, narrow rifts that initiated the splitting apart of Africa and South America. These deposits are younger than those of Europe or the Gulf Coast of the United States, dating from early in the Cretaceous period. Thus the ages of salt deposits along the margins of present-day continents provide a very neat and instructive record of how Pangea broke apart. Of course evaporites are not the only record of continental rifting, but they are especially useful because they are marine deposits, and their ages can usually be bracketed fairly accurately using fossils contained in the more normal sediments interbedded with them.

Not all continental rifts experience the conditions that lead to the formation of evaporite deposits. Nevertheless, even very ancient rifts can usually be identified from the typical succession of rocks that occurs within them and traces their evolution. As the continental crust begins to pull apart, sediments of the kind that would be found in any steep-walled, subsiding valley accumulate: thick deposits of materials washed down from the valley walls, typically rocks such as conglomerates, which are mixtures of relatively coarse rock fragments of a variety of types and sizes cemented together in a finer-grained matrix. Frequently lakes also develop along large rifts, eventually leaving behind patches of relatively fine-grained sediments. The East African Rift valley is a good present-day example of a rift at this stage of development. It is marked by a long string of lakes, the largest of which are Lake

Tanganyika and Lake Nyasa. Volcanic activity is common in rifts, because as the crust of the continent is stretched and thinned, hot material in the mantle wells up and begins to melt. Again, the East African Rift valley, dotted with volcanoes like Mount Kilimanjaro, is a good present-day example. So the signature of a true rift is quite unique, and is likely to include conglomerates, lake deposits, volcanic rocks, and possibly, if the rift evolves sufficiently to allow the influx of seawater, evaporite deposits. Sometimes a rift never succeeds in splitting apart the continental crust, and its great scar is gradually filled in with sediments, leaving little obvious trace at the surface. But if it continues to widen, the sporadic influx of seawater eventually becomes continuous, and the rift becomes a sea or an ocean in its own right—as happened with the Pangean rifts that eventually became the Atlantic Ocean. When this occurs, the unique sequence of sediments that marks the early stages of rift development is preserved only in narrow strips along the opposite margins of the new ocean basin, perhaps separated by thousands of kilometers.

 ## THE WILD WEST

It is fairly obvious from the events chronicled in Chapter 8 that, geologically speaking, eastern North America was a very active place in the Paleozoic era. The Appalachian ranges, formed during the agglomeration of continents into what was eventually to become Pangea, are a legacy of the events that occurred there. But when Pangea broke up again, the story was quite different. Certainly some volcanism occurred along the margin of the rifting continent, but as the rift widened the continent edge was further and further removed from the boundary of the plate, and thus from most geologic activity. The east coast of North America became a passive margin, and the action shifted west.

Along the west coast, from Mexico to Alaska, a great swath of material was added to the North American continent during the Mesozoic. This happened not by the suturing together of large, identifiable continental masses in the way that Africa, Europe, and the Americas were joined to form Pangea. Rather it occurred by the gradual addition of a multitude of small fragments of crustal materials. If there is a present-day example of this process, it might well be the western Pacific. Were all of the island arcs and microcontinents, from Kamchatka and Japan to the Andaman Islands, to be swept up against the Asian landmass, the result might be somewhat like what happened in western North America during the Mesozoic.

Geologists have termed the small fragments of continental material that have been patched onto large continental blocks "exotic," or "suspect," or "displaced" terranes. All three names suggest the unusual nature of these blocks. They were first recognized because they show very abrupt contacts with their surroundings, usually having different ages, containing different fossils, and being made up of different rock types from neighboring portions of the crust. Exotic terranes are not unique to western North America—indeed they have been recognized in the Appalachians and in many other regions. But western North America, where some 200 (!) such fragments have been recognized, is a classic example. Most of these exotic continental pieces were added during the Mesozoic. Figure 9.3 shows just some of the major blocks that have been identified.

Because of the presence of this patchwork of continental fragments, the geology of western North America is very complex in detail. But the basic outline of its history is not difficult to visualize. During most of the Mesozoic there was a zone of subduction along the entire western margin of the continent. Seafloor was being thrust down into the mantle, dragging along behind it any islands or microcontinents that happened to be there. Unlike the ocean crust, however, these materials were not dense enough to sink into the mantle, and they were sutured onto the continent as they reached the western edge of North America. There is evidence that at times there may have been multiple subduction zones more or less parallel to the coast, each with its associated chain of volcanic islands producing crust that would eventually collide with and become attached to North America. The exotic terranes comprise a wide variety of materials, including not only volcanic rocks similar to those of today's island arcs, but also oceanic sediments and sometimes slivers of the ocean crust itself that were caught up between converging plates and thrust eastward onto the continent. The Golden Gate Bridge in San Francisco rests on just such a sliver. In fact, within the city limits of San Francisco you can observe many interesting geologic clues to the events that were

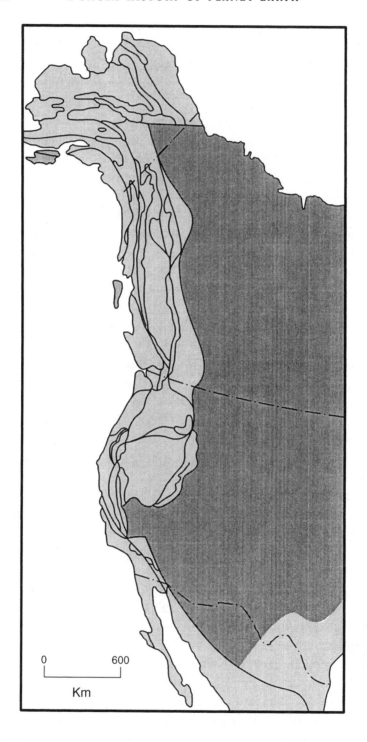

0   600
Km

occurring along the West Coast during the Mesozoic. Geologist Clyde Wahrhaftig from the University of California at Berkeley wrote a small guidebook describing some of these sites; he called it "A Streetcar to Subduction." The official state rock of California is serpentine, a dense, soft, gray-green material typically found in localities where ocean floor has been thrust onto continental crust at a subduction zone; it is a rock well represented among the San Francisco localities described in Wahrhaftig's field guide. It forms through the interaction of water with the rocks of the uppermost mantle, and its presence at subduction zones means that the slabs of oceanic lithosphere that are carried up onto the continents during collisions sometimes are thick enough to contain mantle as well as crustal material.

In addition to the suturing process by which exotic terranes became attached to the continent, western North America also gained new material during the Mesozoic through volcanism. Coupled with the offshore subduction zone was an associated belt of volcanic activity inland, much as occurs in the Andes today. In the Jurassic and Cretaceous periods, this volcanic belt stretched from Alaska to Mexico, creating a broad and spectacular mountain range. Uplift and erosion have worn away most of those volcanic rocks, and only small pockets remain to provide evidence about their ages and nature. But the roots of those great mountains, huge masses of granitic rock that crystallized and solidified deep within the crust beneath the summits of the active volcanoes, are now exposed at the surface. One of the best known of these remnants is the Sierra Nevada, a great patch of red and pink on the geologic map of California. Yosemite National Park, a mecca for tourists and rock climbers, sits at the heart of the Sierra Nevada. But while

←————————

FIGURE 9.3   *Much of western North America is made up of small pieces of "exotic" crustal material that were sutured onto the continent during the Mesozoic. Some of the larger fragments are outlined in the lightly-shaded region of the map at left. Each is bounded by faults (dark lines) and is geologically distinct from its neighbors. Darker shading signifies more ancient crust (see Figure 4.3, page 50). Modified after Figure 16.25 in* The Dynamic Earth, *3rd edition, by B. J. Skinner and S. C. Porter. John Wiley & Sons, 1995. Used with permission.*

the rocks of the Sierras, so loved by the naturalist John Muir and generations of backpackers since, ultimately owe their origin to events in the Mesozoic, the present-day topography is a much more recent construction. By 40 or 45 million years ago, well into the Cenozoic era, there was little trace of a great mountain range, and large rivers drained across what is now the crest of the Sierras to the sea. There is much debate among geologists who study the evolution of the western United States about just when the uplift that created the present range began, but much of it seems to have occurred over about the past five million years. The entire huge block of crust that constitutes the Sierra Nevada has been raised and tilted westward in response to the still-continuing rearrangement of forces in western North America that is related to the changes in the plate boundary—from a subduction zone along the coast, to a transform fault, the San Andreas—that began some thirty million years ago (see Figure 5.6). The spectacular deep valleys and cascading waterfalls of the Sierras are younger still. They were carved from the uplifted granitic block by the waxing and waning glaciers of the past two to three million years. A hundred million years from now the snowcapped volcanic peaks of the Andes will also be long gone, and perhaps there too only the eroded granitic innards of the great volcanoes will remain for future beings to ponder.

 ## THE STORY OF THE MESOZOIC REPTILES

While Pangea fragmented and the physical world began to resemble that of the present day, there were also momentous changes occurring in the biological realm. Gone from the seas were the trilobites and many other animals that had been characteristic of much of the Paleozoic era. On land, forests of seed-bearing plants such as the cycads and ginkgoes (the latter are probably extinct in the wild today, but are popular as shade trees in many regions), as well as the more familiar conifers, became predominant, and late in the era flowering plants became abundant. But the main story of life in the Mesozoic is the story of the reptiles. Like no single group before them, they came to dominate the land, sea, and even the air.

In the previous chapter, it was noted that the reptiles appeared in the latter part of the Paleozoic, evolving from the amphibians. The most important aspect of their development was the invention of an egg that could be laid away from water—the so-called amniotic egg, which takes its name from the membrane (amnion) that surrounds and protects the embryo and the fluids in which it is bathed. This is the familiar egg we boil for breakfast. Chickens, and for that matter all other birds, are descendants of the early reptiles.

The development of the amniotic egg, and also of a type of scaly skin that was a much better protection from desiccation than any that had appeared before, allowed the reptiles to range far and wide on the continents, and to inhabit environments unsuitable for amphibians. Many of them were vegetarians, and the rich array of plant life that by then had colonized the continents provided a ready supply of food. A few early reptiles actually returned to the sea, and developed marine lifestyles, probably as efficient swimmers and predators. Because they were marine, many of these reptiles left good fossil records, and are often displayed in museums. Some of them were huge, reaching the size of present-day whales. It is likely that at least some tales of sea monsters are the products of imaginations stimulated by the fossils of these Mesozoic creatures.

It is on the continents, however, that the main story of the reptiles unfolded. And surprisingly, although the dinosaurs came to be the dominating reptiles later in the era, a quite different group provided the success story of the early Mesozoic. These were reptiles referred to as the "mammal-like" reptiles, and it is from them that true mammals, and eventually humans, evolved. They arose in the latter part of the Paleozoic era, and although they became quite diverse, their numbers were greatly reduced during the mass extinction that ended the era. However, they recovered and flourished again during the Triassic period. They were apparently mostly plant-eaters, although a few were predators and probably devoured amphibians, other reptiles, and eggs. Even though some of the mammal-like reptiles were quite large—the size of a present-day hippopotamus, or larger—their fossils suggest that they were relatively awkward beasts. Indeed, their inefficient locomotion, compared with the dinosaurs, may be one of the main reasons for their demise. Fossil skeletons show that the limbs of the mammal-like reptiles were attached at the sides of their bodies, presumably giving them a sprawling gait compared to most dinosaurs, which had legs attached directly beneath the body.

The mammal-like reptiles laid eggs, but had other characteristics that do not fit our usual concept of a reptile. Some of them probably had hair and whiskers, distinctly mammalian characteristics, and it is likely that some were warm-blooded, or at least had some capability to regulate their body temperature. However, these creatures were nearly all extinct by the end of the Jurassic period, although they did leave us their descendants, the true mammals, which survived but remained generally small and inconspicuous during the remainder of the Mesozoic era. By the Jurassic, the other main branch of the reptilian family, which includes the dinosaurs, had become dominant. But while the mammal-like reptiles are not nearly as well known as the dinosaurs, it is worth remembering that by the end of the Triassic period they had actually been the most important creatures on the land for almost as long as the dinosaurs would subsequently flourish. Although they are not particularly beautiful animals, they are our ancient forebears. To keep your heritage in perspective, take a close look at your Triassic ancestor in Figure 9.4.

Dinosaur fossils were first found in the early part of the nineteenth century, and they were given their name from the Greek words for terrible (dino) lizard (saur). Clearly their imagined appearance in the flesh, even based on these early fossil finds, was fearsome to those who described them. But as we shall see, not all dinosaurs were large, and many probably would not have been any more frightful than common modern animals.

The oldest dinosaur fossils come from fairly early in the Triassic period, about 240 million years ago. Paleontologists have identified two main branches of the dinosaur family based on body structure, in particular the way in which the bones of the pelvis and hip were arranged. The so-called lizard-hipped dinosaurs included both giant carnivores like the famous *Tyrannosaurus rex*, and many less ferocious, plant-eating creatures. The bird-hipped dinosaurs were all herbivores, and included some familiar varieties like *Stegosaurus* and *Triceratops*.

In recent years many of the conventional ideas about dinosaurs have fallen by the wayside, or at least have been strongly chal-

FIGURE 9.4 *A wolf-sized, carnivorous mammal-like reptile that lived early in the Mesozoic era. Drawing courtesy of the Department of Geological Sciences, University of Saskatchewan, Saskatoon, Canada.*

lenged by new evidence. The old concept that dinosaurs were slow-moving, dim-witted, solitary creatures, forced to wade about in swamps because their great bulk could not be supported on land, has been overtaken by modern research. In fact, many dinosaurs were very agile. Apparently they roamed the land easily despite their size, and some were quite social, traveling in herds, building nests, and caring for their young. They may also have been warm-blooded.

When the dinosaurs first appeared in the Triassic, they were small, most no larger than a cat or a small dog. Many were bipedal, capable of moving swiftly on two legs. In contrast to our ancestors, the mammal-like reptiles, even the earliest dinosaurs had legs that were positioned directly beneath their bodies rather than splayed out to the sides. Fossilized tracks of dinosaurs often show footprints spaced far apart and arrayed in a straight line, buttressing the inference from fossils that they had rapid, efficient locomotion.

Many of the small, agile early dinosaurs were flesh-eaters, feeding on other reptiles and amphibians, and perhaps also preying on some of their close relatives among the herbivorous bird-hipped dinosaurs. Many of the latter eventually developed impressive protection against would-be predators: Witness the heavy armor and dangerous-looking spikes on the tail of *Stegosaurus,* or the sharp horns of *Triceratops* (see Figure 9.5). Even in the Mesozoic there was no such thing as a free lunch—the meat-eaters had to work for their nourishment.

FIGURE 9.5   Stegosaurus *(upper) and* Triceratops *(lower) are two of the more familiar ferocious-looking vegetarian dinosaurs. Presumably their horns, armor, and spikes protected them from their meat-eating relatives. The plates on the back of* Stegosaurus *are often shown standing up, rather than flat as depicted here. Drawings courtesy of the Department of Geological Sciences, University of Saskatchewan, Saskatoon, Canada.*

One of the problems of reconstructing the history of the dinosaurs is that the fossil record is relatively spotty. Unlike the marine environment, where expiring organisms are often quickly buried and preserved, the continents are a much less hospitable place for dying animals. Predators and scavengers are likely to pick the bones clean, and also to scatter the skeletons far and wide. Even bones eventually decompose when exposed to the elements on the land surface for long periods of time, wiping out the record of their former owners forever. And streams and rivers may carry animal remains far from their original location, making it difficult to reconstruct the environment in which the creatures lived.

In spite of these problems, a great deal has been learned. Dinosaur eggs and even dinosaur "nests," some with tiny hatchlings as well as eggs, have been discovered. Dinosaur footprints have provided important evidence about the mode of locomotion, and they have also shown that some dinosaurs traveled in groups. There have even been well-worn fossilized trackways found in the western United States, which suggest that at least some of the dinosaurs may have been migratory, traveling seasonally in large herds much as the buffalo did until a few hundred years ago, or caribou still do today in the Arctic.

One of the greatest collections of fossil evidence about dinosaurs ever found is preserved in sediments that were deposited in a large, low-lying area of the western United States, mostly in what are now the states of Utah, Wyoming, and Colorado. The sediments are continental, deposited in the fresh water of lakes and rivers, not in seas. When they were accumulating there was folding, uplift, and volcanism occurring to the west, all ultimately related to the subduction and suturing of exotic terranes to the western margin of the continent that was discussed earlier in this chapter. During part of the Jurassic, large quantities of mud and sand, eroded from this uplifted land to the west, were transported to the lowland areas where the dinosaurs lived, providing the raw material for the mudstones and sandstones that entombed their remains. The preserved fossils are numerous and diverse, and from them much has been learned about the lifestyles of these great beasts. The picture that has emerged is one of a Jurassic equivalent of the East African veldt: a vast area teeming with life, with very large numbers of plant-eating dinosaurs—analogs of the giraffes, zebras, and wildebeests in Africa today—and a small population of predators such as *Tyrannosaurus,* the Jurassic "king of the beasts."

Throughout their long reign on earth—almost 180 million years—the dinosaurs diversified and evolved. Those from the last part of the Cretaceous period would not have recognized their Triassic ancestors. But perhaps surprisingly, nearly all of the major groups persisted right up until the end of the Cretaceous, adding credence to the idea that their extinction was due to a catastrophic, short-lived event rather than to gradual evolutionary change.

One of the more obvious trends in the fossil record of the dinosaurs is change in body size. As already mentioned, the earliest examples are generally small; most of the familiar large beasts of museums and *Jurassic Park* lived in the later parts of the Jurassic and in the Cretaceous. Exactly why this evolution toward larger size occurred is unknown, although there have been many hypotheses, ranging from the idea that larger bodies and longer necks were needed to reach food in large trees, to the suggestion that great size provided protection against predators (or alternatively allowed dominance over others). The largest dinosaurs were truly enormous, with weights estimated to be in the range of 80 to 100 *tons!*

The large size attained by many of these animals has a direct bearing on the debate about whether or not they were warm-blooded, because one consequence of large size is thermal stability. Animals lose heat through the surfaces of their bodies, and the well-known relationship between surface area and volume (which is closely related to weight) accounts for the fact that an iguana would come to thermal equilibrium with its environment much more quickly than would an eighty-ton *Brontosaurus* after a sudden temperature change. For the same reason, it is more difficult for a large animal to get rid of heat generated by metabolic activities than it is for a small animal. As a consequence, it has been suggested that some features of dinosaur anatomy, such as the unusual triangular plates on the back of *Stegosaurus* that are so obvious in Figure 9.5, were used as heat exchangers. The idea is not entirely far-fetched, because it has been discovered from detailed examination of fossils that these bony structures were extensively vascularized. However, *Stegosaurus* was not particularly large as dinosaurs go, and nothing similar to these unusual back fins appears in other dinosaur groups, so their true function is still unclear.

Nevertheless, there is substantial circumstantial evidence that at least some dinosaurs regulated their body temperatures. One of the strongest such clues has to do with the distance between the heart and brain. It is obvious that in many dinosaurs this distance must have been great, as much as several meters, and uphill at that. Thus the blood pressure required to deliver fresh oxygen to the brain cells—without which they would die—would have been quite high. Although the soft body parts of the circulatory systems of dinosaurs are not preserved as fossils, this circumstantial evidence suggests that they were capable of pumping blood efficiently at low pressures from the heart to the lungs to acquire oxygen, and at high pressures from the heart to the brain to deliver it to brain cells. In short, they must have had circulatory systems resembling those of warm-blooded animals with high metabolic rates. However, in spite of the evidence, we can never be sure because—with apologies to Michael Crichton—no one is ever likely to take a dinosaur's temperature.

However striking the reign of the dinosaurs, theirs was far from the only important story of the biological realm during the Mesozoic. The mammal-like reptiles, which actually arose before the Mesozoic started, have already been touched on briefly. Three other groups of organisms are worthy of our attention, although in a book of this size it is impossible to give them (or many other unmentioned groups) the scrutiny they deserve. These are birds, insects, and flowering plants.

THE BIRDS AND THE BEES

We take birds for granted. Nevertheless, no one who has watched a pelican glide along an ocean swell, or a hawk plummet earthward at high speed toward its prey, can help marvel at their mastery of the air. But it was not until late in the Jurassic period, between 140 and 150 million years ago, that the first birds took to the skies. Insects had long since discovered the advantages of flying, and there were airborne reptiles before true birds arose, but all other creatures were either rooted to the earth or inhabited the ocean.

The first flying vertebrates were true reptiles in which one of the fingers of the front limbs became very elongated, providing support for a flap of stretched skin that served as a wing. These were the pterosaurs, literally the "winged lizards." The earliest pterosaurs arose near the end of the Triassic period, some 70 million years before the first known fossils of true birds occur, and they presumably dominated the skies until they were eventually displaced by birds. Like the dinosaurs, some of the pterosaurs became gigantic; the largest fossil discovered is of an individual that had a wingspan of 50 feet or more, larger than many airplanes! These flying reptiles had large, tooth-filled jaws, but their bodies were small and probably without the necessary powerful muscles for sustained wing movement. They must have been expert gliders, not skillful fliers, relying on wind power for their locomotion.

The birds evolved quite separately from the pterosaurs, and have been much more successful in their dominance of the air. They are an example of a common theme in evolution, the more or less parallel development of different types of body structure and function for the same reason—in this case, for flight. Although the fossil record, as always, is not complete enough to determine the evolutionary lineage of the birds in as much detail as one would like, it is better in this case than for many other animal groups. That is because of the unusual preservation in a limestone quarry in southern Germany of *Archaeopteryx,* a fossil that many have called the missing link between dinosaurs and birds. Indeed, had it not been for the superb preservation of these fossils, they might well have been classified as dinosaurs. They have the skull and teeth of a reptile, as well as a bony tail, but in the fine-grained limestone in which these fossils occur there are delicate impressions of feathers and fine details of bone structure that make it clear that *Archaeopteryx* was a bird. All modern birds, from the great condors of the Andes to the tiniest wren in your garden, trace their origin back to the Mesozoic dinosaurs.

Finding *Archaeopteryx* was a very lucky break for paleontologists, because bird fossils are not common. Nevertheless, enough exist to determine that the birds took over from the pterosaurs as masters of the skies during the Cretaceous period. It was during this time that they developed the strong but light, hollow-boned skeletons that characterize modern birds, allowing them to become much more effective fliers than either the pterosaurs or their own early ancestors.

Less conspicuous but nevertheless important denizens of the Mesozoic airways were the insects, which had appeared on the scene much earlier, during the Paleozoic. However, the insects fared quite badly during the mass extinction at the end of the Permian period, and were greatly reduced in numbers at the beginning of the Mesozoic. But they quickly recovered and underwent a remarkable expansion in diversity that continued throughout the era.

Insects enjoy a complex interdependence with plants. Some are scavengers that live on plant debris; others are pests that are truly destructive to some types of plants. Some insects are symbionts, performing essential functions for the very plants that serve as their food source. We often associate insects with flowers or fruit—worms in apples, bees in a blossom—but for much of the Mesozoic there were no flowering plants. Indeed, with only ferns, cycads, ginkgoes, and conifers, the landscape of much of the Mesozoic must have been pleasantly green, but nevertheless drab. In spite of this, careful studies of fossils show that most of the great array of feeding habits and mechanisms of present-day insects were already present even before flowering plants developed. This was a surprising finding, for it had long been assumed that the advent of flowering plants must have been a great stimulus to insect evolution. However, in spite of the fact that many adaptations and reciprocal relationships have developed between specific flowering plants and insects, it appears that on balance it was probably plant rather than insect evolution that benefited most from the development of flowers. Insects were attracted to the flowers, and brought the great reproductive advantage that they carried pollen from one plant to another, inadvertently fertilizing them.

The flowering plants—the so-called angiosperms—developed only 100 million years ago, during the Cretaceous period. However, they very quickly became the dominant type of plant life on land, and remain so today. They populate environments ranging from dry deserts to tropical rain forests, and are found from the equator to the high Arctic. Not all of them produce the kinds of blossoms you would find in a florist shop, but they do all share a key reproductive feature: a seed with a protective coating and surrounded by a ready supply of nourishment. The flowering plants have developed an incredible array of colors, scents and fruits, all through their intricate and reciprocal relationship with the animal world. Not only do insects help to pollinate the angiosperms, but birds and animals disperse the seeds, often over great distances. Without the abundance and variety of the flowering plants that we know today, the world would be a much poorer place.

 ## THE MESOZOIC OCEANS

We close this brief tour of the Mesozoic earth with a few comments about life in the seas. There, as on the land, dramatic changes took place. Perhaps one of the most important occurred late in the era, among the small organisms that populate the uppermost, sunlit portion of the oceans, the plankton. The term plankton is a broad one, designating all of the small plants and animals that float about or weakly propel themselves through the seas. In the Cretaceous period there was a great expansion of plankton that precipitate skeletons or shells composed of two types of mineral: silica and calcium carbonate. Calcium carbonate, as we have already noted, is the main constituent of limestone; silica is SiO_2, the same chemical composition as quartz, and is the main constituent of the rock type chert. This development radically changed the types of sediments that accumulate on the seafloor, because while the organic parts of the plankton decay after the organisms die, their mineralized skeletons often survive and sink to the bottom. For the first time in Earth's long history, very large quantities of silica skeletons, which would eventually harden into chert, began to pile up in parts of the deep sea. Thick deposits of ooze made up of the tiny remains of the calcium carbonate secreting plankton also accumulated as never before. The famous white chalk cliffs of Dover, in the southeast of England, are just one example of the huge quantities of such material that amassed during the Cretaceous; there are many more. In fact, the Cretaceous period takes its name from the Latin word for chalk, *creta*. Just why the calcareous plankton were so prolific during the latter part of the Cretaceous period is not fully understood. Such massive amounts of chalky sediments have never since been deposited over a comparable time period.

The high biological productivity of the Cretaceous oceans also led to ideal conditions for oil accumulation. Oil is formed when organic material trapped in sediments is slowly buried and subjected to increased temperatures and pressures, transforming the organic remains into petroleum. The sediments along the margins of the Tethys Seaway, the tropical east-west ocean that formed when Pangea split apart in the Mesozoic, and that persisted into the Cenozoic, were rich in organic material. Many of today's important oil fields are found in those sediments—in Russia, the Middle East, the Gulf of Mexico, and in Texas and Louisiana.

As already indicated, the Mesozoic appears to have ended with a global catastrophe that wiped out large numbers of plant and animal species, including all of the dinosaurs. Exactly how this sudden mass extinction occurred is not known with certainty, but there is strong evidence that a large impact from space may be at least partly responsible, as will be discussed in the next chapter. Perhaps surprisingly, but fortunately for us, the mammals were apparently little affected in this global crisis. We, the descendants of these survivors, may eventually gather enough evidence from the rocks to determine why.

10

GLOBAL
CATASTROPHES

THE BOUNDARIES BETWEEN eras, periods, and the even finer subdivisions of the geologic timescale are all defined on the basis of abrupt changes in the fossil record. As we have seen in earlier chapters, both the Paleozoic and Mesozoic eras ended with major extinction events in which large fractions of the existing species on earth were eradicated. It is difficult to escape the conclusion that these must have been times of extraordinary conditions for terrestrial life. While these facts have been known by geologists for a long time, and much has been written about the possible reasons for such events, the whole subject of mass extinctions took an incredible turn in 1980. That was when Louis Alvarez, a Nobel Prize-winning physicist from the University of California at Berkeley, working together with geological colleagues—including his son—uncovered evidence for an extraterrestrial cause of the end-of-Cretaceous extinctions.

 ## THE CRETACEOUS-
TERTIARY IMPACT

The smoking gun that they found may at first seem a bit obscure. Alvarez and his coworkers discovered minute amounts of a rare chemical element, iridium, in ocean sediments that had been deposited exactly at the boundary between the Mesozoic and Cenozoic eras. However, although the actual amount of iridium they detected was small, it was more than 100 times greater than was found immediately above or below the boundary.

What does this have to do with an extraterrestrial cause for extinctions? The answer is actually reasonably straightforward. Like its better-known relatives gold and platinum, iridium is a "noble" metal, and is not very reactive. It is also very rare in the earth's crust. Its scarcity is due to the fact that it alloys readily with iron, so that when the earth's core was formed, most of the iridium in our planet was taken up in the sinking molten iron, and it now resides in the core. But the most common types of meteorite that arrive on the earth are the chondrites discussed in Chapter 2, pieces of small asteroids that never went through a planetary core-formation episode. Because of this history, the chondrites retain their full complement of iridium. They have concentrations of this element nearly 10,000 times higher than most parts of the earth's crust. This huge enrichment makes iridium a very sensitive tracer for the input of extraterrestrial material to the earth's surface. Large meteorites are completely vaporized during impact, and as a result their iridium is spread far and wide. On a fairly rapid timescale, it is washed out of the atmosphere by rain and settles to the seafloor, leaving a narrow band with high iridium content in the sediments that are slowly accumulating on the seafloor. Because iridium is relatively inert, the record persists with little disturbance, even over geologic timescales.

The mass extinction at the end of the Mesozoic era marks the boundary between the Cretaceous and Tertiary periods, frequently abbreviated as the "K-T" boundary by geologists (The letter K has long been used to symbolize Cretaceous on geologic maps and diagrams. It comes from the German spelling of Cretaceous, and serves to distinguish this period from others also beginning with a C, such as the Cambrian.) It was in K-T boundary sediments from several different, widely separated localities that Alvarez and his colleagues found large iridium enrichments. Based on their initial measurements, they calculated that an asteroid (or possibly a comet) of about 10 km in diameter could supply the excess iridium, and speculated that the effects of the collision might be responsible for the K-T extinctions. This finding hit the scientific world—and the popular press—like a bombshell. First, what could be more spectacular and science-fiction-like than killing off the dinosaurs with an earth-asteroid collision? Second, although a 10-kilometer-diameter asteroid does not seem especially large given the size of the earth, the consequences of the impact of such an object are mind-boggling.

Theorists have calculated the effects of asteroid impacts in considerable detail. Fortunately, no objects even remotely approaching 10 km in size have been observed impacting the earth; the calculations are based on the results of experiments with much smaller bodies. They are also based on observations of the effects of bombs. Although we would be much better off without them, the weapons of mass destruction have provided occasional scientific benefits.

As for likely causes of extinctions, Alvarez and his colleagues pointed to just one probable aftermath of a large impact: a globe-encircling cloud of dust thrown up by the collision. They noted that the dust would shut out the sun and prevent photosynthesis for several years, killing off plants and much of the food chain that depended on them for nourishment. It would also cause drastic cooling of the darkened surface of our planet. But in addition there would be other, equally serious effects. For example, the shock waves generated as first the asteroid, and then the rocky ejecta thrown up by its collision, passed through the atmosphere would have caused rapid heating and severe atmospheric disturbance. Nitrogen and oxygen, the two most abundant constituents of our atmosphere, would combine to form nitrogen oxides, which in turn would dissolve in precipitation to form nitric-acid rain more corrosive and more widespread than anything yet caused by human activity. The shock-heated atmosphere would also severely desiccate vegetation worldwide, leaving it susceptible to fire and possibly even providing the heat to ignite it. Edward Anders and his colleagues at the University of Chicago have found great abundances of soot particles in the K-T boundary sediments, which they interpret to be the result of widespread, possibly global, wildfires directly related to the impact. There is also evidence in the sediments for tsunamis—huge ocean waves that in principle could reach heights of several kilometers near an oceanic impact. In fact the possible consequences of the K-T impact are so cataclysmic that some geologists have wondered aloud at the fact that so many species actually survived the event. And no wonder—it has been estimated that the energy released would have been at least 10,000 times that of the world's entire nuclear weapons arsenal.

In fairness, it should be pointed out that there are scientists who doubt the impact hypothesis. But the opposition has dwindled as more and more evidence has accumulated over the years since the Alvarez discovery. It has even been possible to identify with some certainty the crater caused by the impact—at Chicxulub, in Yucatán, Mexico. This feature is not now easily recognizable as a crater, because it has been filled in with sediments over the 66 million years that have gone by since the end of the Cretaceous. However, geophysical measurements show very clearly the outline of a circular, buried crater, and drilling in the area has recovered melted and partially melted rock typical of impact craters. Dating of some of this material shows that the impact occurred precisely at the time of the major K-T extinctions. The timing seems too exact to be a mere coincidence.

Because it is not exposed, the exact size of the Chicxulub crater is not known with certainty, but recent studies of the gravitational field in and around the crater suggest that it may be as much as 300 kilometers in diameter. If this is verified by future work, it would indeed be evidence for a gigantic impact, requiring a body much bigger than the original estimate of 10 kilometers in diameter. Certainly its ejecta was spread far and wide. Tiny grains of highly shocked minerals with the distinctive characteristics of bedrock from the Chicxulub area have been found in K-T boundary-layer sediments thousands of kilometers away.

Thus there is a great deal of corroborating evidence for the impact theory. But perhaps the strongest argument remains the excess iridium, which now has been found globally in every complete section through the boundary that has been investigated. It is very difficult to explain this feature except by a sudden, massive input of extraterrestrial material. Whether or not the impact was directly responsible for some or all of the documented K-T extinctions, it seems inescapable that a large extraterrestrial body collided with the earth at the very end of the Cretaceous period some 66 million years ago.

 ## OTHER AGENTS OF EXTINCTION

Although the impact theory has focused much attention on the K-T boundary extinctions, in terms of the number of expiring species this event is dwarfed by the extinctions that occurred at the end of the Paleozoic. Paleontologists had long recognized their severity, but the excitement surrounding the K-T debate has led to a renewed interest in extinction events in general, and those at the Permian-Triassic boundary in particular. Incredibly, about 90 percent of the species that were living in the seas at the end of the Permian did not survive into the Triassic. Life on land was not as diverse then as it was at the end of the Mesozoic, and the fossil record is less complete, but recent work has shown that land-dwellers were also not immune to the devastation. In particular the insects—a group that did not suffer as markedly as others at the K-T boundary—show a precipitous drop in diversity at the Permian-Triassic boundary. However, in spite of the fact that it has been searched for diligently, there is no evidence of an asteroid impact at this time. Other processes must have been responsible. Life on earth, it appears, is in some ways a fragile thing, and can be brought to an end in a variety of ways.

It is useful to remember that the great mass extinctions such as those at the K-T and Permian-Triassic boundaries take place against a background of ongoing extinction that is a normal feature of evolution. What distinguishes the mass extinction events is a dramatic increase in the *rate* at which extinctions take place, and their global nature. The boundaries of the geologic timescale designated by early geologists identify times in earth history when large, rapid, and widespread changes occurred. Theirs was a fairly qualitative assignment: old forms disappeared, new ones took their place, and the boundary was put in between. Modern, more rigorous analyses of the fossil record have employed statistics to examine the rates at which groups of plants and animals have appeared and disappeared. Such studies have highlighted five or six truly major extinction events, and an equal number of smaller ones, since the beginning of the Cambrian period, as shown in Figure 10.1. As might be expected, most of these coincide with the ends of geologic periods, a quantitative affirmation of the earlier observations.

It is clear from Figure 10.1 that mass extinctions are rare events. In fact, by a very large majority, most species that have become extinct during the earth's history have done so as part of the background noise, not in one of the great extinction crises. An important question is, are the causes of the mass extinctions also extraordinary events? For some, the strong evidence for a large impact at the end of the Cretaceous is proof enough that this is the case. Yet at no other major extinction boundary is there a large

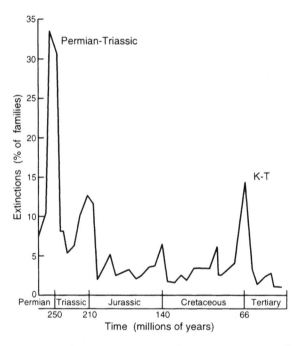

FIGURE 10.1   *The extinction rate of marine organisms (here shown as the percentage of biological families that became extinct) has fluctuated dramatically over the past several hundred million years. The highest rates coincide with the boundaries between geologic periods. Modified after Figure 1 of D. M. Raup and J. J. Sepkoski, Jr., in* Science, *volume 231, page 832. American Association for the Advancement of Science, 1986. Used with permission.*

iridium excess, or any other compelling evidence for a giant impact. This is in spite of the fact that, based on information about impact rates on the moon of the kind shown in Figure 3.1, it is a certainty that other large bodies have struck the earth during its long history. However, it may be that the K-T impact was the largest, and the most lethal, during the past six or seven hundred million years.

One of the key pieces of evidence related to the causes of extinction events is the exact timing of the disappearance of various organisms. Did all species that appear to become extinct at a particular boundary perish simultaneously, or was the mass extinction a drawn-out affair? Did some species actually straggle across the boundary into what we call a different era or period? Unfortunately, such questions are sometimes very difficult to answer, especially for the earlier parts of geologic history. Partly this has to do with the imperfection of the fossil record, and partly it is a result of our frequent inability to determine the precise age of a particular sedimentary rock. It is often impossible to say with certainty whether a particular rock is a few million years older, a few million years younger, or exactly the same age as another rock found halfway around the world, containing different fossils. And when paleontologists believe they have found the last fossil occurrence of some organism, there is always the concern that it may have lived on happily for a considerable period of time in some other region or environment where it was not preserved. That this can indeed happen is well illustrated by what geologists have come to call Lazarus species—organisms that seem to disappear completely from the known fossil record, but then reappear after a long gap, perhaps millions of years. In the biblical story, Lazarus was miraculously raised from the dead by Jesus after just four days, not quite as impressive a time gap as those in the fossil record, but then Lazarus had evidently truly expired. The Lazarus species of the geologic record must have been living *somewhere* during the time when they appear to be missing, but their hiding places have never been found.

In spite of the difficulty of assembling detailed information about just how quickly various mass extinctions proceeded, it does seem from the available evidence that the K-T boundary event occurred much more abruptly than most of the others. This is only to be expected if indeed an impact, which is truly an instantaneous event, played a role in the extinctions. And if the causes of most mass extinctions are seemingly much more mundane than collision with an extraterrestrial object, perhaps something can be learned about their nature by examining what is known about the ongoing background or "normal" extinctions. For these, there are a number of cases from the present, or from the relatively recent fossil record, that provide quite unambiguous information about causes. Climate change—even fairly subtle climate change—is clearly implicated. For example, in the deserts of southern California today there are species of small fish that survive only in a few isolated oases. Eventually, unless protected by man, they will become extinct. A few thousand years ago, in wetter times for this region, these fish flourished in the large lakes of the area—as did the Cahuilla Indians, whose lives also revolved around the lakes.

There is considerable circumstantial evidence that climate change was at the root of some of the major extinction events of the past. Competition, especially competition for food, is another reason for extinction, although it is unlikely to be a dominant one in mass extinctions. It has been argued that competition was responsible for the minor role played by mammals during the Mesozoic. Although poised for evolutionary advances at the beginning of the era, they did not begin to dominate until the dinosaurs disappeared, almost 200 million years later. In historical times, inadvertent or even planned introduction of foreign species by man has often led to decline and sometimes complete disappearance of native populations of both plants and animals due to competition. Significant extinctions of native Australian marsupials, beginning when man first arrived on that continent, are a case in point.

The list of possible agents of mass extinction is quite long. It contains mechanisms ranging from the exotic to the ordinary; some examples are explosion of a nearby supernova, which would have bathed the earth in lethal radiation, the effects of plate tectonics moving continents into and out of favorable climatic belts, and the rise and fall of sea level. Perhaps one of the best ways to explore the possible importance of some of these is to examine recognized extinction events in some detail, and in particular to determine which types of organisms became extinct, and whether there is independent evidence for environmental change of any kind at the same time.

 ## A PRECAMBRIAN EXTINCTION?

The first event recognized by at least some paleontologists as a mass extinction actually occurred in Precambrian time. Its exact timing is uncertain, but it happened near the very end of the Proterozoic era. The organisms most notably involved were the soft-bodied Ediacarans, which were mentioned briefly in Chapter 7, although some species of algae seem to disappear at about the same time. There is considerable debate among paleontologists about just where the Ediacaran animals fit into the general scheme of evolution, and especially over the question of whether or not they are related to the later Burgess Shale–type fauna. Regardless of their relationship to other organisms, however, Ediacaran fossils are widespread in the rocks that were deposited in the shallow seas of the late Precambrian, and are found on most of the present continents. Their relatively good preservation in spite of the fact that they have no mineralized skeletons or shells, and their apparently rapid disappearance, are both puzzling. Indeed, it has been argued that these animals did not experience mass extinction at all, but that their sudden absence from the fossil record is due to a change in the ease of their preservation as fossils. The most frequently invoked possible cause for such a change is an abrupt rise in the number of scavengers, burrowing animals, or oxidizing bacteria—any or all of which would rapidly have destroyed the fragile Ediacaran remains. However, there is no convincing independent evidence that an increase in abundance of any of these organisms actually coincided with the demise of the Ediacarans. In addition, there are many examples of later sediments, without Ediacaran fossils, that in other respects are very similar. These provide no signs of a drastic change in the environment of deposition. Thus the disappearance of this quite diverse fauna, coupled with evidence of the dying out of some types of algae at about the same time, suggest that the late Proterozoic did indeed witness a mass extinction.

If such an event occurred, what was its cause? Sediments from this time period have been examined carefully for excess Ir, which might record an impact, but none has been found. With the available (admittedly scanty) evidence, the best explanation seems to be that the preferred habitat of the Ediacaran animals—shallow water environments—was drastically reduced in amount because of falling sea levels. Analysis of the sediments still preserved from late in Precambrian time suggests that there were repeated cycles of rising and lowering water levels. One of the largest lowerings, also known as regressions, during this time appears to coincide with the extinction of the Ediacarans.

Indeed, it is widely believed that sea level change, particularly the lowering of sea level, was a major factor in many of the extinctions in the geologic record. Because the net effect of weathering and erosion is to grind down hills and mountains, reducing the elevation of the land toward sea level, there are large tracts of the continents at low elevations and probably always have been. In these regions, even relatively modest changes in sea level can have dramatic effects. Biological activity is typically high in shallow seas, and times of high sea level provide abundant habitats for marine life, but when the seas withdraw, many of these organisms become extinct. The total range of sea level fluctuations over the past six hundred million years appears to have been very large, at least 200 meters.

In spite of the possible role of sea level change in extinction events, it is obvious that only marine habitats would be affected. Extinctions that also involve large numbers of land-dwellers (such as the K-T extinctions) cannot be solely due to such changes. Furthermore, not all known sea level fluctuations, even some quite large ones, coincide with large extinctions.

 ## THE PROBLEMS OF QUANTIFYING AND UNDERSTANDING MASS EXTINCTIONS

Geologists and paleontologists attempting to define and understand extinctions are faced with a number of difficulties that are worth examining, if only briefly. It is easy to lose sight of them when we talk glibly of an extinction like the Permian-Triassic extinction killing off 90 percent of marine species. Do we really know this? How good is the evidence?

In fact, we know from examining life in today's oceans that thousands of species would leave no, or at best very sparse, fossil records. This is especially true for the invertebrate animals, for example worms, which are nonetheless numerous and important members of the marine fauna. In the Permian, it is likely that an even smaller fraction of marine organisms was easily fossilized. Furthermore, only some unknown fraction of those organisms that *were* readily preserved as fossils has been discovered and studied. On a dynamic earth, fossil-containing sedimentary rocks may be subducted, metamorphosed, uplifted, and eroded away—and the older the rock, the less likely it is that it has escaped such processes. On the other hand, there are places where sediments containing abundant fossil records of life in the Permian and Triassic have been preserved, and it is from very detailed and careful studies of these localities that the evidence for mass extinction has come. As already noted, modern studies have employed statistical methods, and while the old joke about the statistician who drowned in a river with a mean depth of four inches underscores the fact that statistical investigations sometimes do not tell the whole story, the very large numbers of biological families and genera that have been examined in extinction studies insure that extrapolation to all existing organisms may be a reasonable procedure.

But even the very large database documenting which plants and animals survived across a particular boundary, and which expired, must be treated with caution. An unfortunate but very understandable aspect of paleontology—and indeed most sciences—is specialization. The practical result is that most paleontologists are specialists in fossils from some particular slice of geologic time— the Permian, the Triassic, or even some more restricted portion of the time scale. So the Permian specialist may recognize a particular group of organisms, most members of which disappear from the record at the end of the period. But the Triassic specialist may place the one surviving member of this group into a different Triassic group. In a sense this is a pseudoextinction, at least at the level of the group in question: It "disappears" only because of the classification scheme, not in reality.

Yet another aspect of extinctions that always needs to be examined critically is the question of cause and effect. It is difficult not to believe that the K-T impact had an influence on at least some of the extinctions at the end of the Cretaceous, just because its timing coincides precisely with the boundary as independently established on the basis of fossils. But that is not unequivocal evidence for a causal relationship. The cause and effect problem is even more difficult for the Permian-Triassic and other extinctions for which there is no evidence for an essentially instantaneous catastrophe. We can try to demonstrate that sea level changes, climate changes, or other factors coincide with a particular extinction, but it is only when it can be shown unambiguously that the organisms that became extinct were indeed those most sensitive to such changes that the link can be made with some certainty.

THE GREAT PERMIAN-TRIASSIC CRISIS

The spectacular nature of events at the Cretaceous-Tertiary boundary has tended to obscure the overwhelming importance of the Permian-Triassic extinctions, which saw the end of most of the species then existing in the oceans. The devastation on land was only moderately less extreme. The nature of life on earth was radically changed, and the effects are with us today in the form of all living plants and animals. The causes of this event—or events—are unclear, but it is generally acknowledged that rather severe conditions would have been required to exterminate such a large fraction of life on earth.

The picture that seems to be emerging from Permian-Triassic studies is very different from that of the K-T boundary. The Permian-Triassic record is one of complex extinction patterns in the face of complex and partly interrelated environmental change. No neat, clear-cut culprit has been identified, but much has been learned about the mechanisms of extinction. Nevertheless, the links between cause and effect are still quite tenuous.

The boundary between the Permian and Triassic periods was defined by early geologists on the basis of the great changes they observed in marine fossils. Where are ocean sediments that span the two periods found? Remember that Pangea was being assembled during the Permian, and that by the time of the Permian-Triassic boundary it was essentially a single continent stretching from pole to pole (Figure 8.4). The Atlantic Ocean did not exist. Most of the marine sediments still preserved from that time were deposited along the margins of the Tethys Ocean, the eastern sea that eventually propagated westward, prizing Europe from Africa and North from South America as discussed in Chapter 9. Today these sediments can be found in parts of the southern Alps, in the Middle East, in Pakistan and India, and in China. The record in these regions is complicated by the fact that there was apparently a fairly rapid lowering of sea level near the end of the Permian period, greatly reducing the area of continental shelves on which sediments were deposited. Nevertheless, by careful study of the preserved sequences and correlation from one geographic locality to another, it has been possible to piece together at least some aspects of the great Permian-Triassic crisis.

The simplified and abbreviated summary of the extinction is that on the whole marine organisms were more strongly affected than land-dwellers, that among the marine species those living in shallow waters, and especially those that were anchored to a substrate, seem to have suffered most, and that the extinctions were geographically uneven. There is good evidence that many types of organisms were already in decline before the extinction, and had been for millions of years during the Permian, but it is also generally agreed that there was a very significant increase in extinction rate during the last few million years of the Permian. Depending upon one's viewpoint, a few million years is a short or a very long period of time. However, it seems quite clear that the Permian-Triassic extinctions occurred over a substantially longer time interval than those at the K-T boundary.

What can have caused these selective, uneven, yet devastating extinctions that came closer than any others in the fossil record to wiping out life on earth? The clearest link seems to be with the lowering of sea level at the end of the Permian, which would have greatly reduced shallow-water marine habitats. But this in itself would not have been catastrophic enough to explain the observations, particularly the decline of land-dwelling organisms. Far greater environmental stress was required, and apparently the Permian-Triassic world provided it. For example, the sea level lowering was not an isolated event, but part of several cycles of rising and lowering seas near the end of the Permian period that must have played havoc with life in the shallow waters around the continental margins. When sea level fell, not only were vast shallow-water habitats removed from existence, but also great quantities of organic matter that had been deposited in the sediments, the remains of shallow-water-dwelling organisms, were exposed to the atmosphere. Oxidation of this material produced CO_2, which, as we are frequently reminded these days, is a "greenhouse" gas. CO_2 in the atmosphere traps heat near the earth's surface; as its concentration rises, so does the global temperature. Changes in its concentration can thus have substantial effects on climate.

Furthermore, at the Permian-Triassic boundary, oxidation of organic matter along the continental shelves was not the only source of CO_2. A peculiar and relatively recently recognized category of compounds that occur in sediments of the continental shelf is the so-called gas hydrates. These contain large quantities of gases such as CO_2 and methane, the latter yet another greenhouse gas. They can form and exist only within a narrow range of conditions—for example they form at moderately high pressures, and break down under the normal atmospheric pressure of the earth's surface. The large pressure drop associated with removal of 50 or 100 meters of water from the seas overlying such deposits would have caused them to decompose and release their gases to the atmosphere. Finally, there was an additional source of greenhouse gases: One of the largest known episodes of continental volcanism occurred very close to the Permian-Triassic boundary. These rocks now constitute the Siberian Traps, layer upon layer of lava flows and volcanic debris covering much of central Siberia. (The designation of this and other similar accumulations of basalt flows as "traps" is based on the Swedish word for steps. Especially after erosion, the flat-lying flows often present a steplike appearance.) Many geologists believe that the environmental effects of the Siberian volcanism would have been severe, although their arguments depend quite critically on just how quickly the lava actually erupted. The main effect that could be related to global extinctions would be the release of large amounts of volcanic gases such as SO_2 and CO_2. Interestingly, there is good evidence that several other episodes of large-scale continental volcanism occurred near times of high extinction rates, the most notable being the Deccan Traps, a great pile of lava flows in central and western India that is similar to the Siberian Traps. Age determinations on these rocks indicate that they erupted at about the time of the K-T extinctions. Whether or not these eruption episodes are directly related to the extinctions is unknown, but the coincidence of their timing is intriguing.

Although it is a relatively straightforward matter to date the igneous rocks associated with events such as the eruption of the Deccan or Siberian Traps, technical difficulties make it much more difficult to determine ages for sedimentary rocks, as discussed in Chapter 6. Thus there is considerable uncertainty about the exact timing of some events that may be related to the Permian-Triassic extinctions. For example, because Pangea stretched from pole to pole near the end of the Permian, conditions were appropriate for the formation of polar ice caps, and indeed there is evidence for glaciation in the Permian. The problem is that it is not obvious precisely when this event reached its peak, or if there was associated global cooling. It is possible that the large and rapid sea level fluctuations near the end of the Permian were at least partly the result of waxing and waning of glaciers. Whether or not this was the case, it is clear that the large sea level drop at the end of the Permian was followed quite rapidly by general global warming and a substantial rise in sea level.

Nearly all of the large-scale computer simulations of the response of climate to global warming predict much more variability than we currently experience. Very large differences between average summer and winter temperatures, as well as periodic spells of both intense heat and extreme cold, would be expected. Presumably such fluctuations characterized the warming climate at the end of the Permian. The large Pangean continent would probably have accentuated this tendency, particularly in its interior. Thus even in the absence of an instantaneous global catastrophe such as an asteroid impact, it is not difficult to imagine that with volcanic eruptions, climatic variability, and abrupt sea level changes, the world at the end of the Permian was a particularly harsh place for many forms of life.

THE NATURE OF THE K-T EXTINCTIONS

As seems to occur after all mass extinctions, the groups that survived the Permian-Triassic crisis expanded and diversified greatly in the Mesozoic era. Although there are several times during the era when extinction rates again increased above the background value, the next truly major extinction event in the geologic record occurred at the K-T boundary. As noted earlier, the Cretaceous-Tertiary extinctions have attracted widespread attention because they include the dinosaurs, and also because of the strong evidence for the collision of a large extraterrestrial object with the earth at that time. Indeed, scientific meetings dealing with events at the K-T boundary typically draw an unusually diverse crew of researchers, ranging from biologists to geologists, physicists, and chemists.

The sedimentary record is more complete and more accessible for the K-T than for earlier extinctions, and concentrated examination by scientists from many disciplines has made this by far the most thoroughly studied of the major extinction boundaries. Some of the best records come from drill core samples of sediments from the oceans. The availability of globally distributed samples of sediments deposited far from the influence of the continents has been an important factor in assembling information about the worldwide environmental changes that occurred at the K-T boundary. Unfortunately, because subduction has destroyed all seafloor older than about 200 million years, no equivalent records through the Permian-Triassic boundary exist in today's oceans. All marine samples from this time period come from sediments deposited on the edges of continents now uplifted and exposed, and they are often interrupted by times of nondeposition due to fluctuating sea level.

Paleontologists examining fine-grained sediment cores from the deep sea have a fairly easy time identifying the K-T boundary. It is marked by a very large drop in the number of fossils of plankton, particularly the small organisms that live in the near-surface waters and make their shells from calcium carbonate. Above the boundary these animals eventually regain their prominence, but most of the species are different. In many areas the boundary is recognized by the presence of a thin layer of boundary clay, which is essentially devoid of calcium carbonate fossils but is sandwiched between great thicknesses of younger and older limestone. As far as can be determined, this pattern for the plankton is global, and the change from abundant production of calcium carbonate shells to virtually none is very rapid. The evidence is strong that the cause of at least some of the extinctions was global and fast-acting. The upper, sunlit part of the oceans, where the plankton lived, was clearly strongly affected, and the overall biological activity in the oceans dropped sharply.

On land, in addition to the disappearance of dinosaurs, other changes were also occurring. One that may even be linked to the extinction of the dinosaurs occurred in the plant kingdom. A major part of the evidence comes not from fossils of the plants themselves, but rather from the study of pollen and spores.

Although hay fever sufferers may not agree, pollen is of considerable use to humans as well as to plants. Produced by seed plants and having a range of distinctive morphologies, tiny pollen grains are spread far from their source by wind and animals. They are also quite hardy, resisting decomposition. They accumulate in the slowly depositing sediments of lakes or shallow inland seas, preserving a continuous and often remarkably complete record of the seed-plant flora of the surrounding countryside. In western North America this record has been studied in detail, and near the K-T boundary it shows very great changes that provide valuable information about the world at the end of the Cretaceous.

The most obvious feature of the pollen record is a precipitous drop in the proportion of pollen grains, compared to spores, right at the K-T boundary. Spores come from ferns, and this change implies that there was a sudden decimation of seed plants accompanied by an increase in ferns. Even today, ferns rapidly take over regions where seed plants have been exterminated for one reason or another, only to give way again gradually to the "higher" plants on a longer timescale. In the early Tertiary, the pollen record also shows a gradual increase in the ratio of pollen to spores as seed plants recovered after the K-T crisis. The precise time period over which this rehabilitation occurred is uncertain, but it seems to have been quite short in geologic terms. Some of the Cretaceous seed plants never reappeared, victims of the extinction, but in relatively short order the overall pollen abundance in the sediments regained its previous levels.

An interesting aspect of this record is that the severity of the drop in pollen abundances seems to be much greater in the south than in the north. This has been interpreted by many paleontologists to be the signature of a cooling climate; northern species, already adapted to cold, suffered less. Extinctions among the planktonic organisms in the oceans follow a similar pattern, with tropical forms showing heavier extinctions than those that lived in temperate waters.

The link between the microscopic pollen grains and gigantic reptiles is that the largest dinosaurs were herbivores, and depended on plants for their nourishment. An abrupt decrease in the number and variety of seed plants would have made life very tough for these animals, and for the predators that fed upon them. However, the cause-and-effect questions surrounding the K-T boundary extinctions are still the subject of fierce debate. Were the seed plants and dinosaurs simultaneous victims of a global cataclysm such as an asteroid impact, or did plants suffer first, disrupting the food chain? Is an impact responsible for the cooling climate implied by the patterns of extinctions both on land and in the seas, or is this trend the result of volcanic eruptions in India, or some other, more mundane, cause? Such questions are difficult to answer, but with the wealth of detailed information now being extracted from the sediments that record events at the K-T boundary, they may yet yield to definitive evidence.

It is worth noting that although we have dwelt on the suddenness of the K-T extinctions, for which there is incontrovertible evidence, it also appears that some species, including some of the dinosaurs, may have been in decline for a considerable period of time before the boundary. For these species the fossil record shows that their numbers gradually shrank, and their geographic distributions contracted. However improbable it might seem statistically, many paleontologists have come to believe that the K-T boundary event, probably an asteroid impact, administered a coup de grâce to many species in a world that was already under considerable biological stress.

If indeed there was a final, geologically instantaneous blow that ended the Cretaceous period, it was likely to have been administered either through an impact or the massive volcanism of the Deccan Traps. Certainly all of the most recent precise dating studies have shown that the timing of both the eruptions and the impact are virtually indistinguishable from the timing of the biological extinctions. It is quite possible that *both* these events played a role in the extinctions. Some of the predicted effects of a large impact—global dust clouds, acid rain, giant ocean waves, wildfires—have already been described, and the evidence in the geologic record suggests that all of these phenomena occurred at the K-T boundary.

The Chicxulub crater in Yucatán has exactly the same age as the K-T boundary. This now-buried structure is very large, a testament to the great size—possibly in the range of 20 km in diameter or more—of the object that collided with the earth. But an especially interesting aspect of the crater is that it was excavated at least partly in sedimentary rocks composed of limestone and gypsum—$CaCO_3$ and $CaSO_4$. The intense heating of these materials during impact would have decomposed them, releasing very large quantities of sulfur oxides, as well as carbon dioxide, into the atmosphere. It is well known that the release of SO_2 from industrial sources and volcanoes promotes the formation of aerosols (tiny, suspended droplets of SO_2 plus water) in the atmosphere. When abundant, such aerosol particles form an atmospheric haze that partially blocks incoming sunlight and cools the earth. The effect has been nicely demonstrated, although on a much smaller scale than at the K-T boundary, by the small but measurable global temperature decrease that followed the release of large amounts of sulfur during the eruption of Mount Pinatubo in the Philippines in 1991. Cooler temperatures (by about 0.5°C) were recorded globally for a period of approximately two years, after which they increased again to their previous values. Calculations suggest that the sulfur released in the Chicxulub impact could have resulted in a 10 to 20 percent decrease in sunlight intensity reaching the earth's surface, for a period of about ten years. And if the lavas of the Deccan Traps were also erupting prolifically at this time, spewing additional SO_2 into the atmosphere, the effect would have been further enhanced. Considering that the impact would also have raised a global dust cloud, it seems very likely that the earth would have been in semidarkness, and drastically colder as a result, for at least several years after the impact. Furthermore, SO_2 combined with water makes sulfuric acid. The atmospheric aerosol particles would have been strongly acidic, and as they were gradually removed from the atmosphere by precipitation they would have produced corrosive, acid rain. All in all, the end of the Cretaceous must have been very unpleasant indeed.

With few exceptions, mass extinctions have occurred in every geologic period since the beginning of the Cambrian. We have discussed just two of these, the largest two, in some detail. These were the major extinctions recognized by early geologists when they divided geologic time on the coarsest scale, for they are the basis for the Paleozoic-Mesozoic and Mesozoic-Cenozoic boundaries. We have also discussed, albeit not in such detail, the mass extinction that occurred near the end of the Precambrian, wiping out the enigmatic Ediacaran fauna. Is there a common thread in these sudden and severe interruptions of life on earth? At the moment, the answer seems to be a qualified no. Qualified, because it is possible to implicate climate change in virtually all of these extinctions, although the details of the changes are not necessarily the same in all cases. It does appear that the K-T boundary is unique in the sense that none of the other mass extinction boundaries provide unequivocal evidence for an impact. Perhaps one of the most intriguing coincidences, if that is really what it is, is the fact that the two largest extinctions of the Phanerozoic, the K-T and the Permian-Triassic, each occurs at the same time as a major episode of continental volcanism. Floods of basalt on the continents like those of the Siberian and Deccan Traps are relatively rare in geologic history, yet these two especially large ones coincide with the two most prominent mass extinctions. Is there a cause and effect relationship? Although some geologists have argued that there is, it is really too early to tell. Not all of the flood basalt episodes, not to mention the extinction boundaries, have been dated precisely enough to permit a convincing analysis of their possible correspondence. Furthermore, although it is always possible that some unknown factor is being missed, it does not seem that the effects even of very large-scale volcanism (mainly the release of CO_2 and SO_2 to the atmosphere) would be severe or rapid enough by themselves to lead to massive, global extinctions.

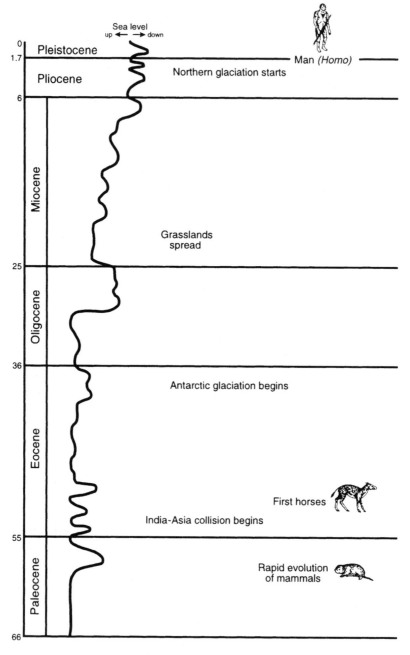

Events of the Cenozoic, with time in millions of years before present. Note that sea level change during the Pleistocene is much simplified; actually many fluctuations have taken place over the past few millions of years, as discussed in Chapter 12.

11

MAMMALS,
MOUNTAINS, AND ICE:
THE CENOZOIC ERA

BY THE STANDARDS of those that have preceded it, the Cenozoic era is very short. Nevertheless, because it is the most recent swath of geologic history, we know a great deal about it. The geographic consequences of its plate tectonic processes are still with us, and the animal and plant life that evolved during the Cenozoic are the familiar fauna and flora of our plains, forests, and seas. In short, the world of the Cenozoic is recognizable and comfortable, easier to deal with than the more distant reaches of geologic time with their unfamiliar creatures and jumbled continents and oceans.

One of the reasons we know so much about the Cenozoic is that the ocean basins contain a complete record of its history in their sediments. All ocean floor older than about 200 million years, and much that is younger, has been subducted, recycled back into the earth's mantle, but a considerable fraction of the seafloor created since the beginning of the Cenozoic era 66 million years ago is still accessible for study. And although we know that the deep sea is not quite the tranquil place it was once thought to be, with a slow rain of sedimentary particles settling to the bottom and accumulating undisturbed, it does nevertheless furnish us with a remarkably complete record of sediments from throughout the era. One of the great accomplishments of the earth sciences in this century has been the organization and execution of the Deep Sea

179

Drilling Project, a large-scale scientific endeavor developed in part to tap that record. Today, more than a quarter of a century after its inception, the project has a different name and is even broader in scope than it was initially, but its goals are still basically the same: to recover long core samples from the ocean floor that will help us to understand the earth's history. Hundreds of geologists from around the world vie for the opportunity to spend two-month stretches at sea on the drilling ship, working around the clock to examine, describe, and analyze the core samples as they are brought on deck. Eventually these cores are delivered to one of three repositories in the United States—at the University of California in La Jolla, at Columbia University in New York, and at Texas A&M University in College Station—where they are further studied and stored under refrigeration for future research. These core "libraries" are an invaluable resource of materials that document the earth's history and are immediately available when new ideas or new methods of analysis are developed. A good example is the question of what happened at the Cretaceous-Tertiary boundary. As told in the previous chapter, the discovery of excess iridium in boundary-layer sediments first led to the suggestion that there was a large impact 66 million years ago. But it has only been possible to demonstrate that this phenomenon is a global feature, and not some local geochemical effect unrelated to impact, through studies of sediment cores from around the world. A large fraction of these has come from the Deep Sea Drilling Project.

Regardless of the exact nature of the events that ended the Mesozoic era, they marked a major turning point in the earth's history. Both on land and in the seas, the course of evolution was altered radically. And although the Cenozoic occupies only about 1½ percent of the earth's history, the operation of plate tectonics, even at its own sedate pace, has changed the physical geography of the world significantly during the era. At its beginning, the Tethys Seaway, discussed in Chapter 9, still existed as a conduit for east-west seawater circulation. Neither the Himalayas nor the Alps had yet appeared. The climate was much warmer than today: There is fossil evidence for subtropical conditions at the latitude of the Arctic Circle. Mammals, although they had been present throughout the Mesozoic era, were still minor players in the biological realm. However, that was soon to change.

One of the truly fascinating things about the Cenozoic is that, because we have so much information, it is possible to link cause and effect with much more certainty than it is for the earlier eras. It is very clear that even the relatively modest plate tectonic shifts in the positions of the continents during the Cenozoic have had strong effects on the course of evolution. These same movements have initiated worldwide climate changes, which in turn have also affected biological evolution. In textbooks on geologic history, the earlier parts of the timescale are often treated by simply reciting the important physical and biological events in separate sections: a mountain building event here, volcanism there; in the oceans such and such organisms flourished, on land this group became extinct and another took over. For the Cenozoic, however, the interconnections between the biological and physical worlds, if not always obvious, are at least much clearer—and it is also apparent that they are very important. Even events that, in the global scheme of things, appear to be quite minor—such as the formation of the Isthmus of Panama some three million years ago, joining the Americas and shutting off east-west ocean circulation between the Atlantic and Pacific—have had major consequences for both climate and the biological realm. While it is unlikely that we will ever know the details of earlier parts of geologic history with the clarity of our view of the Cenozoic, there is a lesson in this knowledge that should not be forgotten when dealing with these earlier times. Even if we can't discern them, the links among plate tectonics, climate, and biological evolution must have been equally strong. Here again, in its broadest sense, uniformitarianism is a useful framework for viewing the past.

  ## THE RISE OF MAMMALS

The Cenozoic era is sometimes referred to as the age of mammals. From aardvarks to elephants, from whales to wombats, and of course including ourselves, the mammals have come to dominate life on earth. They include both tiny creatures like the shrews, which weigh only a few grams, and the gigantic blue whale, probably the largest animal that has ever lived on earth. Although we don't often reflect on it, the lives of man and other mammals are closely intertwined, both in terms of our history and of everyday life. A good deal of our food, and a significant fraction of our clothing, come in one way or another from domesticated mammals. Many important advances in medicine have been made through experimentation using laboratory mammals. And much of the initial exploration of the polar oceans, and of North America and Siberia, was done in search of whales and fur-bearing mammals, respectively, unfortunately often with disastrous consequences for the creatures that were being hunted.

The earliest true mammals that we know about from their fossils lived late in the Triassic period, near the beginning of the Mesozoic era. Even earlier there were mammal-like reptiles, as we have already seen (see Figure 9.4). But for a very long time after their appearance—more than 150 million years—mammals remained small and inconspicuous. The conventional wisdom suggests that this was due to both predation by carnivorous dinosaurs, and competition from all types of dinosaurs. However, after the great K-T extinction eliminated these competitors, the number and diversity of mammals literally exploded. Recent careful studies of the fossil record show that within about ten million years of the K-T boundary, there were some 130 genera of mammals in existence (genera are groups of closely related species), as many as there have been at any time since. Bats, primates, rodents, whales—these and other forerunners of today's animals were already present. Although many species have become extinct since that time, and new species have appeared, the total number of genera has stayed about the same, averaging approximately ninety. This suggests that the initial spurt of mammalian evolution quite rapidly produced a stable population that, at least in terms of number of genera, has not changed drastically since. (It is worth reiterating a point made earlier in this book by pointing out again that phrases like "quite rapidly" need always to be taken in context in geological discussions. In this case the diversification of mammals over *ten million* years is only rapid when considered against the earlier, low-profile history of these animals, which is more than fifteen times as long.)

Mammals are characterized by body hair and the habit of nursing their young. They are also warm-blooded, which has permitted them to adapt to varying climatic conditions more readily than animals that are not, such as the reptiles. However, none of these characteristics is readily fossilized, so the classification of mammals throughout geologic history relies on features of bone structure, particularly the nature of the jaw and the types of teeth. As it turns out, a great deal about a fossil mammal's environment, and especially about its diet, can be learned from its teeth.

Nearly all living mammals bear their young live. The only exception is a rare group, known as monotremes, that lay eggs. These peculiar mammals, which include the duck-billed platypus, are found only in Australia. Unfortunately, the fossil record of the monotremes is almost nonexistent, and the place of the living representatives in the overall scheme of mammalian evolution is poorly known. But, although they are highly specialized, the monotremes have many primitive characteristics and are probably offshoots of the ancient mammal-like reptiles. The most successful of the mammals are the placental mammals, which include ourselves and most familiar animals, both domesticated and wild: dogs, cats, horses, bears, elephants, deer, and many others. The placental mammals give birth only after a long gestation period, and the young (in most cases) are ready to face the world in fairly short order. Marsupials, the second large group of mammals, bear their young at a much earlier stage of development, and must protect them in an external pouch during the earliest part of their lives.

 ## THE MARSUPIALS

The present-day distribution of marsupials provides an interesting example of the interplay between biological evolution and plate tectonics. These animals are most widespread and diversified in Australia and a few neighboring islands (the best known are that continent's kangaroos and koalas), and to a lesser extent in South America. Probably because of the biological cost of the way in which the young are born, the placental mammals seem to have had an edge over marsupials wherever they have been in direct competition. The fossil record shows that the two groups diverged from common ancestors during the Cretaceous period, and the marsupials probably arose initially in South America. Near the end of the Cretaceous—shortly before the remarkable explosion in diversity of the placental mammals—the global climate was warm, and the Antarctic continent was still connected to both South

America and Australia, a lingering remnant of the former southern megacontinent of Gondwanaland (see Figure 9.1). Marsupials migrated from South America across the Antarctic and into Australia. But by early in the Cenozoic, Australia was rifting away from Antarctica and moving northward toward Asia. South America also became an island continent and remained so during much of the Cenozoic, separated from the Antarctic by the Drake Passage and with its land bridge to North America severed, as shown in Figure 11.1.

Particularly in Australia, marsupials were thus able to evolve without undue influence from placental mammals. Relatively quickly they occupied all of the niches that were commandeered by the placentals in other parts of the world. Marsupials that looked and behaved like wolves, cats, and mice, to name just a few, flourished. Some, like the kangaroos, have no similar-looking

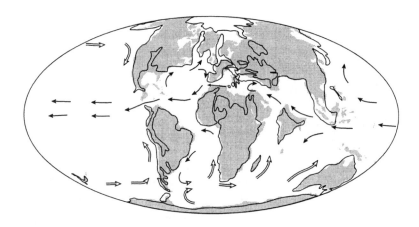

FIGURE 11.1   *The distribution of the continents early in the Cenozoic (approximately 60 million years ago). As for Figure 9.1, present-day continental outlines are shown in gray shading while the coastline of the time is indicated with a heavy line. Note that Australia and South America have just separated from the Antarctic, and a complete circumpolar current, as exists today, has not yet developed. Solid arrows denote warm currents; open arrows indicate cold water movement. Modified after Map 9 in* Atlas of Mesozoic and Cenozoic Coastlines, *by A. G. Smith, D. G. Smith, and B. M. Funnell. Cambridge University Press, 1994. Used with permission.*

counterparts elsewhere, but they occupy equivalent ecological positions to other types of animals, the grazing placental mammals in this case. Unfortunately, many of Australia's marsupials are now endangered because of the introduction of a variety of placental mammals to the continent by man.

Marsupials also diversified widely on the isolated South American continent during the Cenozoic, and in fact, in spite of a large array of coexisting placental mammals, became the main predators there. As in Australia, many forms developed that were the equivalent of placentals elsewhere—for example, wolflike and catlike marsupials are well represented in the fossil record. But when South America was rejoined with North America about three million years ago via the Panamanian isthmus, it was the northern immigrants who won out. Although there are a few success stories, such as the opossum, a marsupial that managed to survive and even to spread northward, for the most part the long-isolated South American mammalian fauna (both marsupial and placental) was decimated by the invading placental mammal predators that invaded across the Isthmus of Panama.

The details of mammalian evolution in South America and Australia could fill several books. They illustrate such things as parallelism in evolution, the effects of competition and climate, and, especially, the role played by plate tectonics in the development of the mammals as they exist today. They show that the slow and inexorable movement of the continents can both create and destroy barriers to animal migration, profoundly influencing the course of evolution.

GRASSES, CLIMATE, AND HORSES

Among the mammals, horses occupy a special, often romantic, place in the human imagination. Talk of horses conjures images of wild mustangs in the American West, of Mongolian horsemen racing across the grasslands of central Asia, of beautiful Arabians cantering across green fields on a misty English morning. Horses have been domesticated for millennia. But what of their earlier history? When did they originate, and how did they evolve? The answers to these questions, following what by now may seem to be a familiar refrain, involve a complex interplay of biological and physical influences. Fortunately, the fossil record of the horses is one of the most complete in all of paleontology, and most of the important changes that occurred between the earliest horses and their present-day descendants are well documented. Their history is truly a textbook case study in evolution, learned by all students of paleontology. There is also a cautionary tale in this classic sequence, however, as has been argued cogently by paleontologist S. J. Gould. True, we can trace the "improvements" in the horse lineage directly from the earliest fossils to the present-day horse. But the path that is outlined in this exercise is only one of many in a maze of branching evolutionary changes, not some inexorably followed evolutionary course. The other branches are now extinct, but there is no way in which this outcome could have been predicted.

It may come as a surprise that at the beginning of the Cenozoic era there were no prairies as we know them—no wide plains with tall grasses waving gently in the wind. The herbivorous dinosaurs of the Mesozoic ate from trees and bushes and other broad-leafed plants. Grasses developed early in the Cenozoic, part of the continuing evolution of the flowering plants, but they occupied fairly restricted environments until about halfway through the era, when widespread grasslands began to appear on the continents. There are various ideas about why this expansion occurred when it did, ranging from the influence of climate to the possibility that it was only when grasses with continuously growing leaves evolved that they could survive the foraging of grazing animals. But regardless of the reason, the spread of grasslands had a significant effect on the evolution of horses, and of other grazing animals as well.

The very earliest known horse fossils come from the Eocene epoch, and they are so different from the modern version that it was not initially realized that there was any relation. Dawn horse, or *Eohippus,* as this animal has been called (although it is properly known as *Hyracotherium*), has been found in both Europe and North America. *Eohippus* was tiny, about the size of a small dog, and apparently lived in wooded areas. Although these animals had hooves, in contrast to modern horses they had several—four hoofed toes on the front feet, three on the rear—and the hooves were padded (see Figure 11.2). *Eohippus* was also pug-nosed by comparison to modern horses, and its teeth reveal that it was a

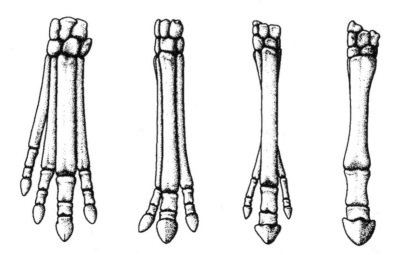

FIGURE 11.2   *The feet of horses have evolved significantly from* **Eohippus** *to the modern horse. Shown here is the change in structure of the rear feet, from four distinct hoofed toes in* **Eohippus** *(left) to one in the modern horse (right). Approximate ages for these four foot configurations are (left to right) Early Eocene, Oligocene, Late Miocene, and modern. Although not shown, there has also been a large increase in size. Reproduced from Figure 319 in* **Vertebrate Paleontology,** *2nd edition, by A. S. Romer. Copyright ©1945 by the University of Chicago. Used with permission of the University of Chicago Press.*

browser that fed on a variety of plants. In fact, although it was completely herbivorous, this little horse had canine teeth—a reminder that many of its predecessors among the Mesozoic mammals were carnivores. The deep, elongated muzzle of today's horses, as we will see, is a direct consequence of tooth and jaw development to cope with a diet of tough and abrasive grasses.

Throughout the Eocene and Oligocene epochs, the descendants of *Eohippus* evolved in a fairly straightforward way that is well documented by their fossils. They became gradually larger; the middle toe, eventually to become the single hoof of the modern horses, became stronger and more prominent, and the grinding surfaces of the teeth became larger, with complex, resistant ridges. But the resemblance to *Eohippus* remained clear. It was only in the Miocene, coincident with the spread of grassy prairies, that abrupt changes took place, resulting in several different lineages of horse evolution, only one of which is still extant: the modern horse. Many other experiments in horse evolution did not make it to the present.

The advertiser's slogan "you are what you eat" might have been coined for horses: Several of the physical characteristics of today's animals are ultimately linked to their diet of grasses. Foremost among the Miocene modifications that led toward present-day horses were changes in the teeth and the shape of the head. Grasses are abrasive, much more difficult to chew and grind than the succulent leaves of tropical trees that were the fodder for some of the horse's ancestors. They contain silica, and can dull even lawnmower blades in relatively short order. The response of the Miocene horses was to develop teeth with much more elaborate and resistant grinding surfaces, and with much larger crowns, at least part of which could grow out of the gums as they were worn down. These changes meant that the head had to be much deeper, and the muzzle longer, to make room for the long rows of grinding teeth along the horse's cheeks. At about the same time, the legs and feet of the ancestors of today's horses became better adapted to rapid running across the spreading grasslands. This occurred through fusion of several of the independent bones in the lower parts of the legs, making them stiffer, and through further emphasis of the central hoofed toe, which by now bore the entire weight of the animal. In place of a foot, the horse has a single toe at the end of its leg, as shown in Figure 11.2.

By the middle or late Miocene, many of the extant horses were at least superficially similar to modern horses. Evolutionary changes have continued, of course, right up to the present, but you would have had no trouble recognizing that the Miocene animals were indeed horses. Based on the fossil record, much of their development seems to have occurred in North America, but by the Pleistocene epoch, the modern horse genus, *Equus,* had spread over much of the world. Then, inexplicably, only eight to ten thousand years ago, horses disappeared from North America. The reason for this extinction is unknown. Some believe it was due to the arrival of man on the continent via the land bridge that joined Alaska and Siberia. Others contend that disease must have exterminated the horses. Whatever the cause, it is a fact that the plains of North America were without these graceful animals for thousands of years, until horses brought from Europe by the early Spanish explorers escaped from their masters and began to repopulate the vast grasslands.

It is clear from the foregoing that many of the familiar features of the modern horse—its speed, the shape of its head, its hooves, and indeed its widespread distribution in the world—are directly or indirectly related to its diet, and its preferred environment, grasslands. But how and why did the grasslands develop when they did? As mentioned earlier, there are competing theories on this question, but only a few are consistent with the evidence. Most of these invoke a change in global climate as an important factor, perhaps the overriding factor. In particular, grasslands expanded rapidly as the climate in continental interiors became cooler and drier.

 ## THE CENOZOIC CLIMATE

Compared to the present, the earth enjoyed a mild climate at the end of the Mesozoic era. This condition persisted into the Cenozoic, and indeed the average temperature apparently increased early in the Eocene, making the time period around 55 million years ago the warmest of the past 70 or 80 million years. But shortly thereafter, the climate cooled precipitously. In spite of some intervening quite long periods of relatively stable temperatures, the earth has been cooling ever since. How do we know this? Temperatures can't be fossilized, but scientific ingenuity has come up with several quite quantitative "paleothermometers" that have been very successful in reconstructing the climate of the past, particularly that of the Cenozoic. Coupled with more qualitative evidence, for example, observations on the latitudinal distribution of certain animals or plants that are known to favor particular temperature ranges, these indicators have provided a very complete record of Cenozoic global temperature fluctuations.

In principle, anything that responds to the ambient temperature in a predictable way, and preserves the response as some sort of fossilized record, could be used as a paleothermometer. Two of the most important such records for Cenozoic temperatures, it turns out, involve such radically different characteristics of the fossil record as the shape of plant leaves and the isotopes of oxygen in limestone.

How could the shape of plant leaves indicate temperature, you ask? Surprisingly, they do so very well. That there is a general relationship between leaf shape and climate has been known since early this century, but in 1978 Jack Wolfe, of the United States Geological Survey, put the relationship on a quantitative footing. Using data for present-day forests in eastern Asia, he showed that there is a remarkable correlation between the mean annual temperature and the shapes of leaves. The particular feature of leaves that seems to be most distinctive in this regard is the nature of the leaf margin (see Figure 11.3). In tropical areas, where temperatures and precipitation are high, leaves tend to be large and have smooth edges, without serrations, and they often have a narrow, elongated tip—sometimes referred to as a drip tip—that facilitates water runoff. In contrast, leaves in cooler regions are typically smaller, narrower, and usually have jagged edges. In today's forests, these characteristics are specific to climatic conditions throughout the globe, even if the actual faunas in different locations are quite distinct. It is a reasonable extrapolation to expect that the same relationship was true at earlier times, and the detailed record of Cenozoic temperatures that paleontologists have constructed from studies of fossilized leaves certainly suggests that this is the case.

The oxygen isotope paleothermometer is a very different kind of temperature indicator, but it tells the same story as the fossil leaves, giving us considerable confidence that our understanding of climate fluctuations in the Cenozoic is basically correct. The method was conceived by Harold Urey, the Nobel Prize–winning chemist mentioned in Chapter 3 for his experiments, conducted together with Stanley Miller, on the origin of life. As already discussed in Chapter 6, isotopes of an element exhibit the same chemical behavior, but differ slightly in mass. As a result, one isotope may be favored over another in a chemical reaction or a process such as evaporation.

A good example of this principle is the effect that evaporation has on the oxygen isotopes in water. As explained earlier in this book, oxygen has three isotopes, of which oxygen 16 is by far the most abundant, making up more than 99 percent of normal oxygen. However, all oxygen also contains small amounts of both oxygen 17 and oxygen 18. A molecule of water is thus likely to be H_2O-16, but it could also be H_2O-17 or H_2O-18. During the pro-

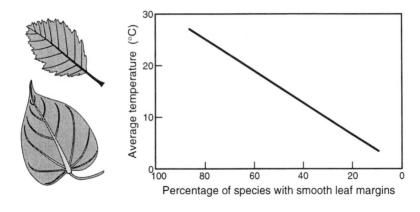

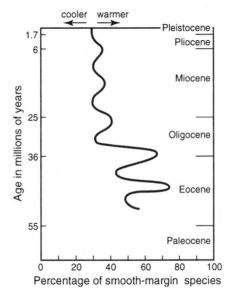

FIGURE 11.3   *Plant leaves may have smooth or jagged edges, as illustrated above. In present-day forests, species with smooth leaf margins dominate where average temperatures are high, as shown in the upper graph, which is based on actual measurements. By applying this relationship to the shapes of fossil leaves from the Pacific Northwest of North America, the temperature history of the Cenozoic has been reconstructed (lower graph). The observed fluctuations, especially the sharp drop near the Eocene-Oligocene boundary, are very similar to those deduced from completely independent evidence such as that shown in Figure 11.4 on page 193. Modified after Figures 1, 2, and 3 of J. A. Wolfe in* American Scientist, *v. 66, p. 695–696. Sigma Xi, 1978.*

cess of evaporation, the lighter water molecules—those containing oxygen 16—have a greater probability of evaporating. The oxygen isotopes in the water are fractionated in the process, water vapor being lighter (containing a higher percentage of oxygen 16) and the remaining liquid becoming heavier (with a larger fraction of oxygen 17 and oxygen 18) as evaporation proceeds.

Urey had been studying the fractionation of isotopes during chemical reactions, and knew that the exact amount of fractionation that occurred was controlled by the temperature at which the reaction occurred. Then he had a truly brilliant insight. He realized that when ocean-dwelling organisms precipitated their calcium carbonate shells, using the dissolved components of seawater as raw material, the relative abundances of the oxygen isotopes in the shells would depend on the temperature of the water. The possibilities were breathtaking. In principle, here was a method that could be used to decipher the history of seawater temperatures through time simply by measuring the oxygen isotopes in the tiny shells of long-dead organisms from ocean sediment cores. Not only that, but because both surface and bottom-dwelling creatures are preserved, it should be possible to learn something about the temperature difference between the surface and bottom waters of the ancient ocean. And furthermore, by analyzing samples of the same age from low and high latitudes, it might be possible to determine the temperature gradient from pole to equator, which, as it turns out, is an important parameter for understanding the global climate.

As is often the case with scientific discoveries, putting Urey's paleothermometer into practice was not quite as simple as it seemed in principle. For example, the snow that eventually forms the polar ice caps is composed of water evaporated from the oceans—a process that, as we have seen, changes the isotope composition of the remaining seawater. Thus during glacial times, the changes in the oxygen isotopes of seawater due to the formation of polar glaciers may be as great as those due to temperature fluctuations. However, in a sense this is just a problem of interpretation. It does not change the fact that the oxygen isotope fluctuations occurred and are permanently recorded in the shells of fossil organisms. Even if the exact temperatures are somewhat uncertain, the timing of temperature shifts can be determined very accurately. Today, studies of oxygen isotope studies are firmly established as one of the most important ways to learn about past climates.

The temperature changes that have been inferred for the Cenozoic from oxygen isotope studies are shown in Figure 11.4. The evidence obtained from examination of leaf shapes corroborates this record very well, particularly the high temperatures of the early Eocene epoch, and the extremely sharp drop in temperatures at the Eocene-Oligocene boundary. The latter coincides with the onset of glaciation in the Antarctic, the development of a polar ice cap. The global cooling during the Eocene has been documented

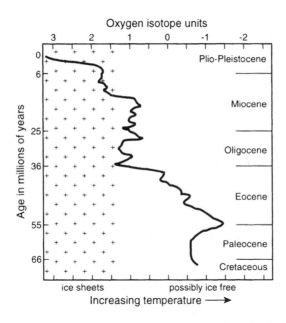

FIGURE 11.4 *Oxygen isotopes in the shells of plankton from Atlantic Ocean deep-sea cores can be interpreted in terms of past sea-water temperatures. Note the sharp drops near the Eocene-Oligocene boundary, and over the past few million years, probably indicating the onset of permanent glaciation in the southern and northern polar regions, respectively. In the units used here, oxygen isotope values above about 1 appear to correspond to times of significant global glaciation. Modified after Figure 1 of K. G. Miller, R. G. Fairbanks, and G. S. Mountain in* Paleoceanography, *volume 2, page 3. American Geophysical Union, 1987.*

in great detail in western North America based on fossil leaf studies. The evidence shows that not only did the average temperature decrease, but the seasonal temperature range got larger, and the climate became drier. Forests declined, and grasslands flourished. Horses and other grazing animals evolved in parallel.

Although many factors influence climate, it appears that the two largest, most abrupt fluctuations during the Cenozoic, both of them temperature decreases, occurred at least in part through the influence of plate tectonics on ocean circulation. At the end of the Mesozoic and into the early Cenozoic, the same land connections between Australia, Antarctica, and South America that allowed the marsupials to spread into Australia, also prevented the circumpolar flow of water around the Antarctic continent. Instead, cold waters flowed north into the Indian, Pacific, and Atlantic Oceans, mixed with tropical water, and the return flow of warm water to the south kept the polar region relatively warm—and ice free (see Figure 11.1). However, as Australia, and eventually South America, separated from the Antarctic continent in the Cenozoic, the cold polar waters could flow around the continent, as they do now, isolating it from warmer water masses to the north (Figure 11.5). The Antarctic became colder, and a permanent ice cap developed, a feature that itself had a noticeable cooling effect on the global climate. The onset of glaciation in the Antarctic as deduced from other evidence appears to coincide closely with the sharp drop in seawater temperature near the Eocene-Oligocene boundary (shown in Figure 11.4) that is implied by the oxygen isotopes.

The second abrupt temperature decrease illustrated in Figure 11.4 occurred three to four million years ago, a time when another plate tectonic change affected ocean circulation. It was at about this time that the gap between South and North America was closed by formation of the Isthmus of Panama, blocking equatorial waters of the Atlantic from flowing westward into the Pacific, as they had done previously. Instead, the Gulf Stream became stronger, carrying relatively warm water northward along the eastern coast of North America. Under the already cool climatic conditions, this warm current provided abundant moisture for precipitation in northern regions, and fairly rapidly resulted in development of the North Polar ice cap, again with the effect of lowering temperatures in other parts of the globe. Like the sharp

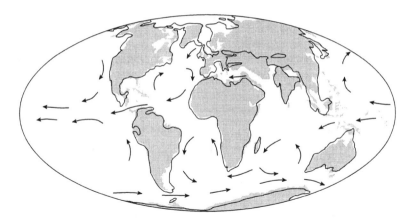

FIGURE 11.5 *By the Oligocene, the Antarctic was isolated from other continents and a circumpolar current had developed. This map shows the continental configuration approximately 30 million years ago. As for other diagrams of this type, dark lines indicate the continental edge of the time, while gray shading denotes today's continental shapes. Modified after Map 5 in* Atlas of Mesozoic and Cenozoic Coastlines, *by A. G. Smith, D. G. Smith, and B. M. Funnell. Cambridge University Press, 1994. Used with permission.*

drop in the Eocene, this temperature change also had recognizable effects in the biological realm. A clear message from the Cenozoic geologic record is that climate, plate tectonics, and evolution are inextricably linked.

 MOUNTAINS IN EUROPE AND ASIA

A few geologists I know are mountain climbers, but many people who pursue that pastime are probably unaware of the extent to which they are indebted to plate tectonics for their pleasure. Mount Everest and the Matterhorn, to name just two famous climbing peaks, owe their existence ultimately to the breakup of Gondwanaland, and the slow northward movement of the continental fragments to eventual collision with more northern landmasses. The collisions that made both of these mountains occurred in the Cenozoic. In fact, the Cenozoic era could as easily be called the age of mountains as the age of mammals. Over its relatively short 66-million-year time span an amazing amount of mountain building went on.

A glance at a relief map of the world shows that there is an essentially continuous band of mountains stretching from Spain and northern Africa across Europe and the Middle East, to India and China and even Indonesia. In sketch form, this belt is shown in Figure 11.6. Although the individual ranges of this huge mountainous region have different names—Pyrenees, Alps, Caucasus, Pamirs, Himalayas, and others—they were all formed as the continents of the former Gondwanaland collided with Europe and Asia.

During the Cretaceous period, toward the close of the Mesozoic era, the Tethys Ocean lay to the south of Europe and Asia. Along its margins were warm, shallow, biologically productive waters; the sediments that accumulated there incorporated much organic

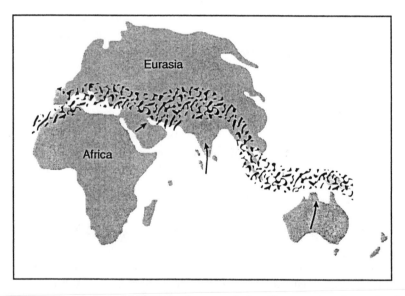

FIGURE 11.6 *As the continents of the former Gondwanaland moved northward and collided with Europe and Asia, belts of mountains were formed (patterned region) stretching from northwestern Africa and the Alps to the Himalayas and Indonesia. Modified after Figure 15.29 in* Evolution of the Earth, *5th edition, by R. H. Dott, Jr., and D. R. Prothero. McGraw-Hill, 1994. Used with permission.*

material, and they now supply a considerable portion of the world's petroleum needs. But the days of the Tethys were numbered. In response to seafloor spreading in the Atlantic and the southern oceans around Antarctica, Africa swung northward toward Europe. In a general sense, the Alps and their associated mountains throughout Europe, North Africa, and the eastern Mediterranean region can be said to have formed by the collision between the African block and Eurasia. However, nature is rarely simple and neat, and the formation of the Alps is no exception. Even geologists who have studied these mountains for a lifetime are puzzled by some aspects of their geology. But by stepping back and using a broad brush, it is possible to understand the general outline.

In the region south of Europe, the Tethys had never been a wide sea. As described in Chapter 9, it had formed during the Mesozoic by the progressive rifting of Pangea from the east. In addition to the main African plate, there were apparently several microplates, small continental fragments, within the Tethys Ocean between Africa and Europe, remnants of the process that had split these two continental masses apart. The initial stages in the formation of the Alps occurred early in the Cenozoic era as these microplates collided with Europe, compressing and thrusting up onto the northern continent pieces of themselves, as well as fragments of the intervening seafloor and sediments of the continental shelf. Two of these microcontinents constitute the regions we now know as Italy and Spain.

Formation of the Alpine belt continued as the African plate pressed inexorably northward toward Eurasia, closing up the Tethys Seaway. Because the northern parts of the Atlantic Ocean were opening up due to seafloor spreading at about the same time, there was also east-west movement between the African and European plates, rotating and grinding both the microplates and the edges of the continents, and complicating considerably the task of geologists trying to unravel the history of the resulting mountain ranges.

Collisions between continents usually occur on protracted timescales, even in geologic terms. This is partly because of the slow rate of plate movement, and also because continental edges are typically irregular, and even if straight, they are unlikely to be parallel during collision. The collision process that formed the

Alps and related mountains in Europe, North Africa, and the eastern Mediterranean is no exception. It lasted for most of the Cenozoic, ceasing less than ten million years ago. It was a complex, multifaceted event, but because of its relative youth, the mountains that were formed have provided considerable insight into the kinds of processes that must have occurred during similar collisions that produced older, now much-eroded mountain ranges such as the Appalachians in the eastern United States.

One of the most spectacular aspects of Alpine geology, at least for geologists, is the presence of features known as nappes. These structures are mute testament to the tremendous compressive forces that occur when plates collide. A good way to visualize what a nappe is, without actually seeing one, is to think about picking up a rectangular carpet by its center, so that part of it hangs vertically, folded to double thickness, while the rest remains flat on the floor. Then let go, with a slight shove to one side or the other, so that most of the carpet remains folded. Nappes are analogous huge folds in solid rock, now lying more or less horizontally, with lengths that are many times their thickness. These giant folds have often been thrust for distances of tens of kilometers over other rock formations of completely different origin. In ancient collisional mountain belts, where, because of erosion and metamorphism, only small fragments of such features remain, their relationship to surrounding rocks can be very puzzling. Even in the Alps, nappes are often partially destroyed by erosion, but they can usually be traced from peak to peak across the eroded valleys.

There is still subduction going on in the Mediterranean region today, the result of the general northward movement of Africa and continued jostling between the Eurasian and African plates. In this process, part of the Mediterranean seafloor is being thrust under Europe. The distinctive clue is an arc of volcanic islands above the subduction region. The active volcanoes of the islands off the northern coast of Sicily, such as Stromboli and Vulcano, and also the Greek islands of the southern Aegean Sea, all owe their existence to the same process that forms the volcanoes of the Aleutians, Indonesia, and the Andes: Water, trapped in the subducting slab and dragged down into the earth's interior, lowers the melting point of the already hot mantle and initiates melting.

If the African plate continues to move northward, the Mediterranean will eventually suffer the same fate as its forerunner, the Tethys Ocean, and eventually disappear as Africa and Europe become welded together. Actually, there is evidence from cores drilled by the Deep Sea Drilling Project that there have already been periods in the past when the Mediterranean Sea ceased to exist, although not because of the suturing together of Europe and Africa. Buried under the normal sediments on the Mediterranean floor are great thicknesses of salt deposits, in places more than a kilometer thick. The age of these deposits is about six million years; apparently at about that time the Strait of Gibraltar was temporarily closed, preventing exchange of water with the Atlantic, and the Mediterranean simply evaporated away, leaving behind only the salt that had been dissolved in its waters.

Because the salt content of seawater is well known, it is a simple matter to calculate how much salt should have been deposited when the Mediterranean dried up. However, the observed thickness is much greater than could possibly have been formed during a single evaporation episode. It appears that the barrier to influx of Atlantic water was a fragile one that was periodically breached (presumably producing spectacular waterfalls into the Mediterranean basin in the process), and that the thick salt deposits result from multiple cycles of filling and evaporation.

Far to the east of the Mediterranean and the Alps lies another remarkable topographic feature of the earth: the Tibetan plateau and the Himalayas. This region is the largest and highest upland area on the planet, home to the Abominable Snowman and (in better times) the Dalai Lama. It too results from the breakup of Gondwanaland and continent-to-continent collision, in this case collision between India and Asia.

India is part of the same lithospheric plate that carries the continent of Australia, as is obvious from Figure 5.2, but in the breakup of Gondwanaland it separated from Antarctica much sooner than did Australia. It had already moved away from the Antarctic before marsupials migrated there from South America, for example, and no marsupial fossils have been found in India. For tens of millions of years, India literally raced (geologically speaking) northward toward Asia, at speeds of more than ten centimeters per year. In the plate tectonic framework, this requires that the intervening seafloor was subducted, and indeed there is evidence of this in the rocks of the Himalayas. Metamorphosed but recognizable remnants of volcanic arcs are found there, the telltale signatures of a subduction zone along the southern margin of Asia.

About 55 million years ago—the precise timing is a matter of debate—the great collision began. For a long time prior to this, small fragments of crust—exotic terranes in the parlance used earlier—had been swept up against Asia on the subducting seafloor. They are now part of the Tibetan plateau. But the first contact with India seems to have been in what is now its northwestern corner, and then the continent slowly rotated counterclockwise, closing the remaining expanse of the Tethys Seaway like a gigantic jaw. In the process, small pieces of seafloor that didn't get subducted were thrust up onto land, and can now be found in Tibet. Some of the highest peaks of the Himalayas are partly composed of ocean sediments from the margins of the Tethys Seaway, scraped off and thrust onto Asia during the collision.

Although the chronology of collision can be constrained to some degree by dating rocks from the Himalayas, using radioactive decay methods as described in Chapter 6, it is not always possible to determine if these rocks were formed during the collision, or if the ages actually reflect earlier events, or if the geologic clocks were reset by metamorphism. Fortunately, there are additional clues to the timing. Because India had been an isolated island continent since late in the Mesozoic era, even before the beginning of the age of mammals, the arrival of diverse continental mammal groups that had evolved in Asia is a distinctive feature in the Indian fossil record. This occurred about 45 million years ago, indicating that by this time a suitable migration route between Asia and India had been established over land.

In spite of the fact that the India-Asia collision began more than 50 million years ago, the uplift leading to the present-day Himalayas is much more recent. As already mentioned, continental collisions are protracted affairs, and there was a long period as India rotated to the northeast, closing up the seas along its entire northern margin, before suturing of the continents was complete. The first indication that there were substantial mountain ranges developing comes from sediments that were deposited in the Arabian Sea, the Bay of Bengal, and on the Indian continent itself. Mountains produce quite distinctive sediments. Regardless of the rock types that occur, steep mountain slopes and rapid runoff mean that the eroded debris shed from them is characteristically coarse-grained. Such sediments first show up in the ocean off the mouths of the Ganges and Indus—the principal rivers draining the Himalayas—near the middle of the Miocene, approximately 20 million years ago. Similar sediments appear at about the same time in the deposits of shallow seas that covered parts of the Indian continent at that time.

The rate at which India was moving north slowed sharply when it began to collide with Asia, but compression between the two continents continues to this day. The forces involved in such collisions are almost unimaginable. The crust of India, made up of typically low-density continental rocks, cannot be subducted deep into the mantle—it is too buoyant. But as it crumpled against Asia, it did try to follow the seafloor that preceded it down the subduction zone, some of it sliding under Asia and producing continental crust almost twice as thick as exists anywhere else in the world. Inevitably, the great stresses accompanying this process have caused the crust to break and fracture. Much of the uplift of the Himalayas over the past few million years has occurred as slivers of crust, with nowhere else to go, were squeezed upward along steeply dipping faults as India pressed relentlessly northward against Asia. The process is sporadic, not continuous, with sudden lurches—and very large earthquakes—occurring when the stresses across a fracture become too great and two adjacent rock bodies slide past one another.

But the immediate vicinity of the Himalayas is not the only place where destructive earthquakes related to the collision of India with Asia occur. As India drove northward, the crust thickened, parts of it were shoved down under the Asian crust, and other parts were thrust upward along faults. But this didn't accommodate all of the movement. In addition, parts of Asia were squeezed and rotated eastward, out of the way of the still-moving Indian continent. Most of the movement occurred along east-west trending faults, and it is still going on because India is still moving north. The consequences are evident even thousands of kilometers away. Earthquakes responsible for the loss of tens of thousands of lives in China have occurred along faults related to the collision.

And Lake Baikal in southern Siberia, the largest freshwater lake in the world, occupies a rift in the crust that was probably formed as Asia squeezed and rotated out of India's way.

Before leaving the subject of Cenozoic mountain building, it is worth touching briefly on the effect such events can have on world climate. We have already seen that by changing oceanic circulation patterns, the movement of continents can influence climate, as happened when Gondwanaland broke up, Antarctica was isolated by a circumpolar current, and a permanent ice cap developed. Mountains, on the other hand, affect atmospheric circulation. Sometimes they act as simple barriers to surface wind flow, and may strongly influence the distribution of precipitation. Such is the case along the west coast of North America, where wet Pacific air is forced upward over ranges such as the Sierra Nevada in California, causing it to lose most of its moisture. As a result, not far east of ski areas where winter snowfalls of three to four meters or more are common, lies the dry desert of Death Valley. The Himalayas and the Tibetan plateau have an even more dramatic effect, for they play a large roll in the Indian monsoon, part of an atmospheric circulation system that influences climate patterns throughout the globe. As summer approaches, the sun heats the high Tibetan plateau, and the air above it, initiating changes in the atmospheric circulation pattern that draw moist tropical air, and welcome precipitation, from the south and west to the Indian sub-continent. Careful examination of fossil records from the region has shown that the strong seasonal monsoons that characterize the present climate of this area developed only after elevation of the Himalayas and Tibetan plateau.

THE COOLING EARTH

We have already seen that various indicators of climate, such as leaf margins and oxygen isotopes, show that temperatures on the earth have been dropping since early in the Eocene epoch (Figure 11.4). Eventually it became cool enough in high latitude regions for winter snows to remain year-round, and the earth had entered a new Ice Age. Such periods have occurred sporadically throughout the earth's history, but they are actually quite rare. They leave a distinctive record in the rocks, in the form of gravelly sediments scraped up by the ice and deposited along the margins of the glaciers, or the annual varves described in Chapter 4, or as glacial scratches and grooves left in the bedrock by the passing ice. The Cenozoic glacial episode is sometimes referred to as the Pleistocene Ice Age, because many of the obvious effects of the great ice sheets that have periodically covered parts of Europe and North America were produced during the Pleistocene epoch. But in fact this is a misnomer. The record in the rocks shows that the permanent Antarctic ice cap formed as early as about 35 million years ago, and permanent northern glaciers were present nearly 3 million years ago, well before the Pleistocene began.

Cenozoic glaciation transformed the landscape in much of the Northern Hemisphere. Indirectly, it provided us with one of our most valuable geologic resources, deposits of sand and gravel. It is responsible for some spectacular scenery, and it also gave us the myriad of lakes that dot the northern parts of Russia, Europe, Canada, and the United States. The course of evolution—in particular the evolution of man—has been strongly influenced by the waxing and waning of polar glaciers during the Cenozoic. And in spite of the fact that we now live in an interglacial—a temporary warm period—there is no reason to suspect that the current glacial episode is at an end. A short 15,000 years ago, the locations where many modern European and North American cities now stand were buried under great thicknesses of ice, and in the future advancing ice sheets may cover them again. In the next chapter, we will complete our journey through geologic time by examining current thinking about glaciation, climate change, and the history and effects of the "Pleistocene" glaciation.

12

THE GREAT ICE AGE

THE INHABITANTS OF Bombay or Riyadh might dispute it, but the earth is currently in the grip of a glacial episode. True, the present is an interval of relative warmth, an interglacial period, but for the past several million years the planet has been colder, on average, than it has over much of its history. Today, there are continent-sized ice caps in both the Northern and Southern Hemispheres. Only 300 kilometers from the equator, on Mount Kilimanjaro, there is a permanent, 5-kilometer-wide glacier. The causes of the periodic deep-freeze events that have affected the earth appear to be quite complex and are not well understood in spite of decades of study. But the details of the most recent Ice Age, the one we are still in, are becoming increasingly well documented. Such things as variations in ice volume, changes in sea level, the response of land vegetation to changing climate, and even the actual temperature fluctuations that have occurred over the past few million years are quite well known. The story revealed by this information is fascinating, doubly so because human evolution has occurred during this period, and has been strongly affected by the varying climate. The record shows that local and even global climatic regimes have sometimes changed rapidly on timescales that are short even by human standards. It suggests that small changes in factors that do not themselves seem very important can, through interactions with other influences and because of feedback mechanisms, produce significant shifts in climate. Many of these sudden shifts have had well-documented effects on the course of human civilization. Even without human-influenced climate change, we should be prepared for large variations in the future. As a famous American banker is reputed to have predicted about the stock market, the climate, in all probability, will fluctuate.

 # THE RECOGNITION OF ICE AGES

It is interesting to speculate whether or not continental-scale glaciation would have been recognized from the rock record were it not for the fact that the earth today sustains numerous glaciers. Early in the nineteenth century, several European scientists realized that the glaciers they could examine and study in the Alps and elsewhere must have been much more extensive at some time in the past. They came to this conclusion after observing that there are deposits in areas far removed from the present-day ice that are much like those found at the margins of active glaciers. In 1795 James Hutton, the Scottish geologist who was the first to articulate the principle of uniformitarianism, had speculated in print that strange "erratic" boulders near Geneva must have been carried to their present location and deposited there by glaciers. The nearest glaciers as he wrote were tens of kilometers distant from the boulders. (Although Hutton did not know it, the boulders actually came from even farther afield.) But the man who is most associated with the general acceptance of continent-scale glaciation is Louis Agassiz, a Swiss geologist who gathered information about glacial deposits throughout Europe and eventually also in North America. Initially a skeptic, Agassiz became convinced by the evidence that much of northern Europe had been buried under a thick ice cover in the past. Very few of his contemporaries agreed—in fact, several senior scientists of the day, giving what they thought was friendly advice to a young scientist gone astray, suggested that he go back to his studies of fossil fish, which had already earned him a reputation as a first-rate paleontologist before he turned to glaciers. But Agassiz was not deterred. He and his assistants climbed unscaled mountains to observe glaciers better, measured their flow rates, and studied the moraines (piles of gravel and boulders) deposited at their margins. His evidence was so compelling that eventually he won over even the doubters. In 1847 Agassiz moved to the United States to become a member of the faculty of Harvard University, and in his travels in the Northeast of his new country he found abundant signs of glacial activity. Agassiz was ecstatic. He was an enthusiastic speaker and a devoted teacher who exhorted his students to learn not only from books but also from nature, and although he continued his work in paleontology, it was his popular lectures on continental ice sheets that drew widespread attention. In recognition of his contributions to this field, a great glacial lake that formed along the edge of the retreating ice in North America some 12,000 years ago was named Lake Agassiz (see Figure 12.3). It was centered approximately where the present Lake Winnipeg stands, in the Canadian province of Manitoba, and at its maximum extent covered an area more than four times larger than Lake Superior.

The work of Agassiz and others showed that northern Europe, most of Britain, Canada, and the northern United States had been buried beneath several *kilometers* of ice in the not-too-distant geological past (see Figure 12.1). These early workers did not have the benefit of radioactive chronometers or other modern analytical tools to date and determine the characteristics of the glacial period, and they deduced only that there had once been great, continuous glaciers, possibly extending from the North Pole to the inhabited latitudes of Europe and North America. They pointed to Greenland as an analog of the conditions they envisioned in the past for the environs of Edinburgh or Montreal. Today we know from the details of the geologic record that the Ice Age of the past few million years has been much more complex. The idea that there had been a single ice sheet advancing from the pole is certainly incorrect; indeed, there were many centers of ice accumulation in North America, Europe, and Asia, from which ice flowed in all directions. We know also that there have been multiple advances and retreats of the ice, at remarkably regular intervals, and that the climate in high latitudes has accordingly fluctuated from one not unlike today's to one characterized by severe cold. In the Northern Hemisphere, vegetation zones—tundra near the ice in the north, then spruce forests, then deciduous forests further south—marched up and down the continents like so many armies

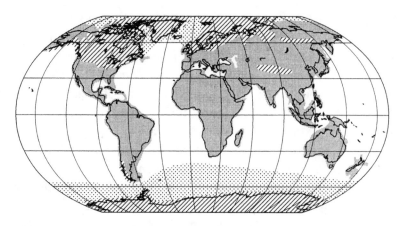

FIGURE 12.1 *Map of the world showing the extent of ice coverage at the peak of the last glaciation. The extension of dry land beyond the present-day shoreline at the glacial peak, when sea level was about 120 meters lower than today, is shown in gray shading. Note the land bridge between Asia and North America. Thick glacial ice over the continents is indicated by the hatched pattern; the dotted pattern denotes sea-ice over water.*

surging back and forth in battle as the glaciers waxed and waned. Near the equator the changes were much less marked, but at midlatitudes they were striking. The most recent advance of the glaciers peaked only some 20,000 years ago, with the ice extending south of the Great Lakes in North America and covering Scandinavia, Northern Europe, parts of northern Russia, and much of Britain. Nearly a third of the present land area of the planet bore a mantle of ice. Similar conditions may prevail in the not too distant future, because if the past is any guide, our present warm period will soon end. However, as we don't yet know exactly what precipitates glaciation, that possibility is still largely speculation.

THE GLACIAL RECORD ON LAND

As we saw in the previous chapter, the oxygen isotopes in seawater are influenced by both the temperature of the oceans and the amount of glacial ice on the continents. Fortunately, both cooling temperatures and ice formation change the isotope values in the same direction, so that even if the two effects cannot be untangled in detail, the timing of the glacial fluctuations is very well documented. The sudden changes that occurred about 35 million years ago, near the Eocene-Oligocene boundary, and again over the past few million years (see Figure 11.4), have been interpreted as reflecting the onset and rapid growth of continental ice caps in the Antarctic and Arctic regions, respectively.

It is the recent glaciation of the Northern Hemisphere that is best documented. The oxygen isotope records from the deep sea indicate that it began in earnest close to three million years ago, and other evidence corroborates that conclusion. Although glacial geology has long had an enthusiastic following, over the past few decades a very large and international effort has been mounted to try to work out in detail the climate changes that have occurred during this "Great Ice Age," and to understand their causes. If there is to be any hope of predicting future climate, and the perturbations that may be introduced by the activities of man, it is clearly important to understand the recent past.

In the tradition of earlier work, the geologists who first began to study the glacial deposits of Europe and North America tried to organize their observations in terms of geologic sequences. They were still without the benefit of the radioactive clocks of later years, and had to rely on correlation of various glacial features from one locality to another in order to build up a relative time framework. In most places each phase of advancing glaciers has scraped away the evidence of previous glaciation, but in a few locations they were able to find repeated layers of glacial sediments on which soils had been developed during the ice-free interglacial times, only to be buried again under the jumbled debris of the next glacial advance. In Europe and North America the detailed record of these events seemed to indicate that there had been four or five discrete periods when glaciers covered much of the Northern Hemisphere. Each of these was named following the time-honored geologic tradition of using a locality name where the rock record is especially well-preserved. In contrast to the earlier parts of the geologic timescale, however, different names have been retained in Europe and North America for what are probably the same time periods, partly because fossils are sparse in glacial sediments and it has therefore been difficult to correlate individual episodes across the Atlantic. In North America, the most recent event is termed the Wisconsin glaciation; in most of Europe the equivalent glaciation is called the Wechselian. It began about 130,000 years ago, and its end is conventionally placed at 10,000 years ago, although, as shown by the oxygen isotope record (see Figure 12.4), the ice volume began to decrease abruptly shortly after the glacial maximum some 20,000 years ago, and has continued its decline essentially up to the present. We now know that there have been many more glacial episodes during the present Ice Age than the four or five identified by early workers; as many as twenty cycles spanning the past two million years have been recognized in deep-sea cores, which, unlike the glacial sediments on the continents, contain an essentially continuous record of the changing climate over long time periods.

Working out the sequence of the last few glacial advances and retreats on land has been a difficult and painstaking process. It has required detailed mapping of the deposits left by the glaciers, and because there were apparently many local variations in the way the ice behaved—perhaps advancing in one region while it was retreating in another—it is not always easy to correlate events across large areas. The dating techniques described in Chapter 6 have helped, but even these are not a panacea because the most useful method, carbon 14 dating, is restricted to the past 50,000 years or so, which covers less than half of the most recent glacial cycle. For most other techniques, the perennial problem of sediment dating that was described in Chapter 6 applies: There are usually no components in the glacial sediments that were formed at the time they were deposited. This means that the ages measured for, say, pebbles in a glacial moraine have nothing to do with the timing of the glaciation. Instead they date the formation of the parent rocks. But geologists have been inventive, and a variety of other methods have been found to measure the ages of glacial features. In the western United States, for example, volcanoes of the Cascades Range, such as Mount Saint Helens, have erupted periodically over the past few million years, and the ash clouds from the larger eruptions left thin bands of volcanic material in glacial deposits throughout the West and Midwest. These can be dated by conventional techniques, and even traced back to the parent volcano. It has also been discovered that the same bombardment by cosmic rays that produces carbon 14 in the atmosphere reaches the earth's surface, although in much weaker form, and produces radioactive isotopes in rocks. When freshly scoured bedrock is uncovered after having been buried in thick ice, it is exposed to this radiation, and the quantity of radioactive isotopes that builds up in such samples is a measure of the time since the rock was stripped of its glacial cover. Several other innovative methods have been developed, so that an accurate chronology of glacial events has gradually been constructed.

But what, exactly, are the features in the rock record that have been mapped and dated in order to work out the extent and timing of glaciation in the past? The most obvious are the deposits of glacial sediments such as moraines, till, and erratic boulders, all of which can be observed near present-day glaciers. Erratics, as the name suggests, are unusual rocks that bear no resemblance to the bedrock of the countryside in which they reside: for example, large chunks of granite in a region where nothing but limestone is to be found. Early observers, realizing that such boulders must originate far from their present locations, thought that they must have been transported by the waters of the biblical flood. James Hutton, as has already been mentioned, was one of the first to suggest that they were carried by glaciers. Till is a general term for the unsorted mixture of material—ranging from fine-grained soil and clay to gravel and boulders—carried and deposited by glaciers. Till is common throughout much of northern Europe, the northern United States, and Canada, and, especially where it has been reworked by running water, is the source of the most economically valuable by-product of the glaciers: sand and gravel for construction. Moraines are simply till piled up in distinctive mounds, usually along the edges of a glacier. Their scale provides a clue to the immensity of the glaciers: Most of Long Island in New York, for example, is a moraine. These ice-built features are also responsible for the pleasant, rolling countryside common in much of the Great Lakes region of North America.

Most of the till deposited by the great continental ice sheets came from quite distant sources. The glaciers, flowing outward from the regions of maximum thickness under the force of their own weight, scoured away existing soil and even some of the bedrock. Picking up gravel and rock fragments at their base, they acted like gigantic sheets of sandpaper pushing across the landscape, smoothing out the topography in some places and accentuating it in others where softer rock units were worn away and harder ones less affected. The resulting scratch and scour marks are still obvious today, at scales from centimeters to tens of kilometers or more. Using air photographs and satellite pictures, geologists have mapped out the orientations of both these marks and long trails of till, some stretching for hundreds of kilometers, to determine the directions of ice flow and locate the thickest regions of ice accumulation. Such studies have shown that there were multiple centers even within the single continent of North America. As the glaciers melted back toward these centers in each interglacial, their load of sand, gravel, rock "flour," and erratic boulders was dropped, leaving many parts of the glaciated countryside buried under an accumulation of till.

An interesting footnote to the study of glacial tills is that in a few places in the United States they have been found to contain diamonds. In the states lying immediately south of the Great Lakes, near the southern limit of the last great ice sheets, some eighty diamonds of a variety of sizes have been recovered from the glacial sediments. The first of these were discovered over a century ago, and it was fairly soon recognized that they must have been carried from somewhere far to the north by the glaciers. Diamonds are formed deep within the earth, at depths of 200 kilometers or more, and are brought to the surface in rare volcanic magmas called kimberlites. The diamond-bearing tills suggest that there are kimberlites somewhere to the north of the Great Lakes, probably in the vicinity of Hudson Bay or James Bay. Although there has been a considerable amount of exploration for these deposits, none has yet been found. Somewhere in the Canadian tundra there are diamond mines waiting to be discovered.

Ice is not a particularly dense material, but a three-kilometer-thick glacier nevertheless adds a tremendous weight to the earth's crust. Just as removal of material by erosion in mountainous regions causes the crust to rise (as discussed in Chapter 4), addition of weight makes it sink. The surface rocks of central Greenland today are depressed approximately to sea level by the weight of the ice cap. Ice has roughly one-third the density of rocks in the mantle, so that adding three kilometers of ice to the crust should cause it to sink about one kilometer into the underlying plastic mantle in compensation. In reality, the effect may not be so great, because the mantle, although yielding, is very viscous. Response to the changing mass of glacial ice, both sinking and uplift, are slow. However, in Scandinavia, in North America around Hudson Bay, and in other regions of thick ice accumulation, the crust was significantly depressed at the peak of ice thickness. As the ice retreated during the present interglacial period, the crust again rose, but slowly. In places this uplift, termed glacial rebound, is still going on. Although sea level also rose as the great ice sheets melted, in most places the land was rebounding even more quickly, and has continued to rise after disappearance of the ice, often resulting in a series of raised beaches, old shorelines that are now well above sea level. Like other glacial features, these have been well mapped, and they show nicely where the thickest ice occurred because these are the regions that were most strongly depressed, and that have therefore most strongly rebounded. In many cases the raised beaches have also been dated, using the carbon 14 method on pieces of wood or other organic material they contain, and from this information the rate of uplift can be calculated. A classic example, shown in Figure 12.2, is from Scandinavia. Raised beaches and other features have been used to construct bull's-eye-like contours of the crustal uplift that has occurred since the ice melted away some 10,000 years ago, and that is still going on.

Two additional effects of the most recent glaciation that have shaped the surface of the land are worthy of mention. One is the widespread presence of loess, a fine-grained, windblown sediment that is found over a significant fraction of the continents, and the other is the occurrence of bizarre landscapes that point to gigantic floods.

The origins of loess are complex, but all deposits of this peculiar type of sediment that have been carefully studied appear to have originated during the height of glacial periods. Some loess is simply the rock flour ground up by the glaciers and redistributed by winds, but some has other origins. During the glacial periods the interiors of continents, particularly in mid- and even low latitudes, were cooler and more arid than today, in many cases with less vegetation. Wind systems may also have been more vigorous. The result

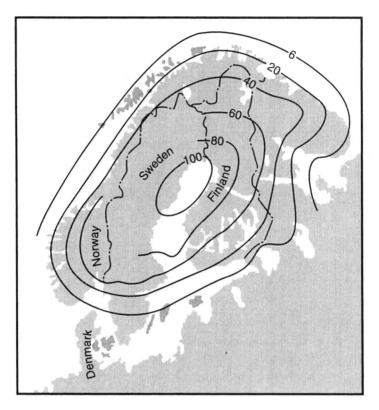

FIGURE 12.2   *Raised beachlines and other evidence indicate that the crust in Scandinavia has rebounded substantially since the ice of the last glacial maximum melted. The contours show uplift in meters, and illustrate clearly where the ice accumulation was greatest. Modified after Figure 19-30 in* Earth, *4th edition, by F. Press and R. Siever. W. H. Freeman and Co., 1986.*

was more effective erosion and transport of fine-grained material. We know that the increase in atmospheric dust was global in extent, because studies of ice-cores drilled both in the Antarctic and in Greenland show that layers corresponding to the glacial maxima are "dustier" than other parts of the record. The most famous loess deposits occur in China, where man-made caves have been carved into the several-hundred-meter-thick accumulations for habitation. The details of the finely layered structure of the loess record fluctuations in the glacial climate, much as do the sediments of the deep sea that are discussed in the next section.

As the ice sheets of the Northern Hemisphere retreated after the maximum of the Wisconsin glaciation, meltwater lakes such as Lake Agassiz developed along their southern margins. Their drainage was constantly changing as the ice withdrew (and sometimes, during short cold periods, readvanced), as the crust rebounded in response to the removal of the glaciers, and as river channels cut through rock barriers. Occasionally, deep lakes broke through ice dams or other obstructions into new drainage routes, with resulting catastrophic flooding. One of the best documented of such occurrences involves a glacial lake in what is now the eastern part of the state of Washington in the western United States. Here, between 12,000 and 16,000 years ago, a large lake, Lake Missoula, went through repeated cycles of filling up, then breaching an ice barrier to release enormous volumes of water westward across the basalts of the Columbia plateau and into the Columbia River. In the process the floods gouged canyons into the bedrock, carved out huge potholes, and left gigantic ripple structures more than 5 meters in height and spaced over 100 meters apart. The region affected by these glacial floods is known as the Channeled Scablands—a name that gives an inkling of the uniqueness of the topography. Its features long puzzled geologists, especially those who were so wedded to Hutton's ideas about uniformitarianism that they could not conceive of periodic catastrophic events shaping the landscape, but eventually their origins were understood. Many other superfloods associated with the waning ice caps have been recognized, both in Eurasia and North America. Probably the largest of these occurred about 8,000 years ago when Lake Agassiz, by then joined with other lakes along the margin of the decaying Canadian ice sheet (see Figure 12.3), suddenly broke through the ice and emptied northward into Hudson Bay. Although the rate at which this process occurred is not known, the volume of water involved was enormous: It is estimated that the level of the *entire ocean* rose by between 20 and 40 centimeters as a result of this flood!

 ## GLACIAL RECORDS IN THE DEEP SEA—AND IN THE ICE ITSELF

As implied earlier, it is in the oceans that the most continuous record of glacial climate changes is preserved. Even in the tropics,

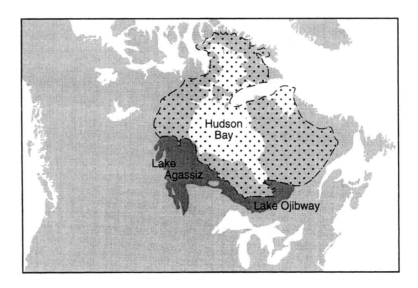

FIGURE 12.3   *Map showing the location of the waning North American ice sheet (dotted pattern) as it stood about 8,500 years ago. A huge continuous lake (dark gray shading), incorporating the waters of Lake Agassiz and other smaller lakes, was dammed along the southern margin of the ice. About 8,000 years ago these waters broke through the crumbling glacier and emptied into the North Atlantic Ocean through the Hudson Strait. Ice and lake locations based on information from* Ice Age Earth, *by A. G. Dawson. Routledge, 1992.*

far from the direct influence of the polar ice caps, the sediments exhibit features that are closely linked to the cycles of advance and retreat of the glaciers. In fact, it has only been since long cores of ocean sediment became available for study that it has been possible to decipher the true details of the Great Ice Age. Although there are many clues to the glacial climate locked in the sediments, quite possibly the most valuable is the record of seawater's oxygen isotope composition.

The ocean-dwelling organisms that construct their shells from calcium carbonate lock in the oxygen isotope characteristics of the surrounding seawater when they do so, thus recording a signal that reflects both water temperature and the amount of water that is tied up in glacial ice. Data such as those shown in Figure 11.4 indicate that the past few million years have been a time of steadily decreasing ocean volume and temperatures. But as shown in Figure 12.4, the record is much more complex if the scale is blown up to show the details of the past few hundred thousand years.

There are several remarkable aspects to this figure. The first is its regularity: The oxygen isotopes in seawater, and, by implication, the glacial cycles, have varied in an amazingly systematic way over the past half-million years. Only five glacial periods are shown here, but if the record is extended back in time to almost three million years, the pattern is similar. It indicates that there were periodic alternations of cold and warm periods. The length of the cycles shown in Figure 12.4 is roughly 100,000 years. For earlier parts of the record, the cycles appear to have been somewhat shorter, but in spite of this it is clear that something affects the earth's climate in a very regular way. There is a beat to the glacial periods that must be governed by an influence that varies in a similar fashion. Based on current knowledge, the only explanation that seems feasible is that the cause is external to the earth, and probably has to do with the amount of heat energy received from the sun.

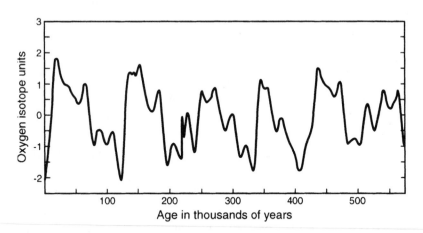

FIGURE 12.4 *Regular changes of the oxygen isotope composition in shells of bottom-living organisms reflect the changing ocean temperature and ice volume of the past 600,000 years. Positive values on this graph indicate cold, glacial periods, while negative ones signify interglacials. The record from deep-sea cores extends much further back in time than is illustrated here, and shows numerous additional glacial-interglacial fluctuations.*

Another important observation that can be made from Figure 12.4 is that the last five cold, glacial periods have been considerably longer in duration than the interglacials, and that the onset of warm periods usually followed the time of maximum ice cover very abruptly. If the present interglacial follows the pattern of the last few such intervals, we do not have long to wait before the climate deteriorates again, in spite of the fact that the Wisconsin glacial maximum occurred only 20,000 years ago. The reasons for the rapid onset and short duration of the interglacial periods are not known.

Up to this point the discussion has assumed that the oxygen isotope changes reliably document variations in global temperature and ice cover. But do they? Are there ways to check this conclusion independently? One of the most convincing pieces of collaborating evidence comes from a source that at first might seem unlikely: tropical corals. Coral reefs grow very close to sea level. Raise sea level by a few meters, and the corals die—but new ones grow on top of them, closer to the new ocean surface. By continuously growing upward, the reefs keep pace as water levels rise, and they are thus very good indicators of past sea level. In places like the Caribbean such reefs have been cored and studied, and their ages determined using the carbon 14 method and other techniques. Corals that lived near the sea surface thousands of years ago are now found at depths of tens of meters, buried in the reef by their progeny. By measuring their ages, and the depth at which they now reside, a record of past ocean levels has been constructed (Figure 12.5). It illustrates that the most recent low stand of seawater occurred at the same time the oxygen isotope data indicate a glacial maximum, about 20,000 years ago. It also indicates that there have been two or three times over the past 20,000 years when sea level rose very rapidly, almost instantaneously in geologic terms, presumably in response to especially rapid melting of the ice sheets. Over the past 20,000 years the oceans have risen by more than 110 meters, covering very large areas that were dry land at the peak of the glacial episode.

Although the oxygen isotope composition of past seawater has probably provided more detailed information about the workings of the glacial cycles than any other single piece of evidence, it is not the only clue contained in ocean sediments. The fossil record of the plankton, for example, shows that—as one might expect—

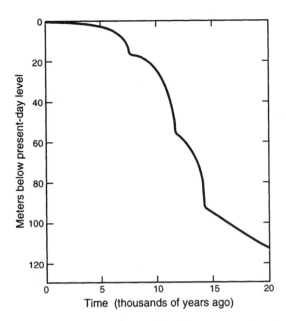

FIGURE 12.5 *Sea level has risen by almost 120 meters over the past 20,000 years as the continental ice sheets melted. This graph, based on studies of now-submerged corals, shows that there were at least three intervals when sea level rose very rapidly—near 14,000 years ago, 11,500 years ago, and again at about 7,600 years ago. Melting of the ice caps on Greenland and Antarctica would raise sea level by an additional 65 or 70 meters. Modified after Figure 3 of P. Blanchon and J. Shaw in* Geology, *v. 23, page 5. Geological Society of America, 1995.*

the range over which warm-water species existed contracted during the glacial episodes and expanded during the interglacials. Some species, less tolerant of cold temperatures, became extinct during glacial periods. Grains of pollen carried to the oceans by rivers and the wind and preserved in the sediments have also provided a rich store of information about climatic conditions during the glacial cycles. Studies of the pollen preserved in sediment cores from along the west coast of North America and elsewhere indicate that the mix of vegetation at any particular location changed in step with the cycles defined by the oxygen isotopes. Taken together, the various clues from ocean sediments have provided a much clearer picture of climate variations during the Great Ice Age than could ever be obtained from the record on land alone. And in recent times, a new source of data has been added to the arsenal of those studying glaciation: the ice itself. In both the Antarctic and in Greenland, deep cores have been drilled through the ice cap. Even in the very cold Antarctic, the winter-to-summer changes in temperature and precipitation are sufficient to leave annual layers in the accumulating ice, so that the cores can be dated very precisely by painstaking counting of the layers. The deepest cores extend back through two glacial cycles, to about 250,000 years. Oxygen isotopes in the ice provide a complement to the seawater record. In addition, the ice cores contain other information that cannot be obtained from ocean sediments. Evidence of the "dustiness" of the atmosphere has already been mentioned, but perhaps the most valuable clue bears on the actual composition of the atmosphere. When it forms, the ice traps small bubbles of atmospheric gas, and by carefully extracting it from the ice core samples, geochemists have been able to reconstruct past variations in atmospheric composition. One particularly interesting result of such studies is the observation that there were fluctuations in the concentrations of two greenhouse gases with the potential to cause temperature changes, carbon dioxide and methane. These gases varied systematically along with the oxygen isotope cycles, their atmospheric concentrations being much lower during the cold periods, and higher during the interglacials. Whether this is a cause or an effect is still hotly debated.

WHAT CAUSES GLOBAL GLACIATION?

If we knew the answer to this question with any certainty, many scientists currently working on this topic would have to look for other problems on which to focus their creative energies. To be fair, there is actually quite a good understanding of the general set of conditions that is necessary, or at least sufficient, to plunge the earth into a cold episode. What is less clear is the nature of the trigger that has caused the earth to fluctuate between warm and cold periods with such regularity over the past several million years. There is no shortage of ideas on this question, but no single mechanism has emerged as the clear favorite. What does seem obvious from the accumulating evidence is that there must be complex interactions and feedbacks among a number of different factors that individually do not have the capability to initiate the observed changes, but working in concert, do. Not much variation is required to tip the balance. On a global scale, the temperature differences between glacial and nonglacial times may be only a few degrees Celsius, ten at the very most.

A characteristic of glaciation that has long been recognized, but that has taken on great significance since the discovery of plate tectonics, is that polar ice caps cannot form on the open sea. Even if other factors cause the planet to cool, large-scale glaciation can only occur if there is land in high latitudes. The fact that the large Antarctic continent is situated squarely over the South Pole is undoubtedly the reason that its present ice cap formed earlier than that of the Northern Hemisphere, and persists as a major feature even during warm interglacial periods such as today. For other times in the past when there is evidence for extensive glaciation, continental reconstructions invariably show large land masses near the poles. For example, the southern continents that made up Gondwanaland—India, Australia, Africa, Antarctica, and South America—all contain glacial till, scoured bedrock and other indications of ice cover dating to late Paleozoic time, between about 250 and 300 million years ago. This is just when Gondwanaland was situated over the South Pole.

Continents at high latitudes are a necessity for ice ages, but so are two additional factors: a ready supply of snow, and cold temperatures, especially in summer. Paradoxically, the first of these conditions requires moderately warm ocean waters, at least in mid-latitudes, in order to promote evaporation and supply atmospheric moisture for precipitation in the polar regions. As already mentioned in Chapter 11, one of the theories that has been advanced for the initiation of Northern Hemisphere glaciation is that the formation of the Isthmus of Panama some three million years ago diverted warm Atlantic Ocean water northward and increased precipitation in eastern Canada, Greenland, and Scandinavia—three of the main centers of thick ice accumulation. But even increased snowfall won't start a glacial episode if it all melts away during the summer. Temperatures must be cool enough that there is net accumulation.

The temperature at any location on the earth's surface is controlled by a multitude of factors, but on a global scale the important ones are how much energy is received from the sun, and how much is trapped by the oceans and atmosphere rather than being radiated back into space. Long before it was known that the earth had experienced regular glacial advances and retreats, mathematicians and astronomers had shown that the amount of energy received from the sun at any particular location must have varied in a regular way in the past because of the earth's orbital characteristics. The astronomical theory of glaciation is generally attributed to Milutin Milankovitch, a Yugoslavian mathematician who lived from 1879 to 1958, and indeed he is the man who developed the ideas in the most detail and essentially in their present form. But even before Milankovitch's work, others had suggested that glaciation might be the result of orbital changes that resulted in less solar energy falling on the earth. Perhaps most notable among them was the Scottish intellectual James Croll, who first published his ideas in 1864. Croll's is an interesting story: When his paper on glaciation appeared, this self-taught man was working as a janitor, one of the several professions he followed while studying and writing on a variety of subjects. Eventually his talent was recognized, and he was appointed to the Scottish Geological Survey, but as time passed his ideas about the Ice Age were given less and less credence. Various arguments were mounted against them, primary among them the fact that the changes in solar energy received by the earth because of orbital variations seemed to be much too small to account for significant climate change.

Long after Croll had died and his theories about glaciation had been all but forgotten, Milankovitch began his mathematical investigations of the earth's orbital variations, and their effect on climate. His initial work was published in the 1920s, all the calculations done by hand—a daunting task. Milankovitch laboriously computed the variations in the solar energy received in the Northern Hemisphere over the past 600,000 years. In his calculations (and in others that have been done since) he assumed that the sun's output remained constant over this time period. This is an aspect of Milankovitch's theory that is subject to debate, because even small changes in the sun's energy production could have significant consequences for the earth. But even with constant output,

Milankovitch had to consider three different ways in which the amount of incident energy could vary: by small, regular changes in the angle of the earth's tilt toward the sun, by slight changes in the shape of the earth's elliptical orbit, which bring it closer to or farther from the sun at the extremes of the orbit, and by a slow rotation of the earth's orbit that gradually shifts the time of our closest approach to the sun from winter to summer and back again. These variations all operate on different timescales, sometimes enhancing and at other times canceling one another, but the important point is that they are regular. Like Croll's earlier work, Milankovitch's calculations generated much excitement when they were initially published, and there was a flurry of activity attempting to relate known glacial deposits to Milankovitch's cycles. However, again following the fate of Croll's ideas, Milankovitch's work faded from prominence as arguments were raised against it. But that situation changed suddenly when geologists developed the capability to collect and study deep-sea sediment cores. As we have seen, sediments deposited over the past several millions of years contain remarkably regular variations in a number of their characteristics, all related to the glacial cycles.

Milankovitch's work has been redone by computer in recent years, and some refinements made, but the story remains basically the same. And although the argument persists that the changes in received solar energy caused by these cycles are themselves not large enough to initiate or end glacial periods, the fact that mathematical simulations of past climate that incorporate the Milankovitch variations agree quite well with the actual geologic record has convinced most scientists working in this field that astronomical factors are somehow at work, perhaps acting as a trigger, the proverbial straw that breaks the camel's back, when other conditions are just right.

The Milankovitch cycles show how solar energy received by the earth has varied with time, but how much of that energy is retained? This is an even more complex problem than calculating the orbital variations, because it depends, among other things, on the distribution of land and sea, the nature of the land surface, and the composition of the atmosphere. Seawater, for example, absorbs most of the sun's energy that it receives, but ice, or deserts, reflect large fractions of it. Continental ice caps therefore provide a positive feedback, reflecting solar energy and cooling the planet further just by their presence. But the ice caps occur at high latitudes where the amount of incident energy in a given area is much less than it is in the Tropics, and thus the cooling effect of high-latitude glaciers would be countered by a land distribution with large oceans and few continents at low latitudes. However, changes in the distribution of the continents are very slow, and while they must affect the sensitivity of the earth to variations in other parameters, they cannot explain the rapid oscillations between glacial and interglacial conditions of the Great Ice Age.

The composition of the atmosphere, on the other hand, does undergo substantial changes over short timescales. Analyses of the gas trapped in ice from Greenland and Antarctica, as already alluded to, indicate that both carbon dioxide and methane in the atmosphere have varied in step with the climate during glacial cycles. Both of these greenhouse gases block the heat that radiates from the earth's surface from escaping into space, and ice cores show that the concentrations of both increased during warm periods and decreased during cold ones. However, close examination of the timing of these changes indicates that in most cycles they seem to lag slightly behind the temperature changes. If this is verified by additional work, it would suggest that they are the result of the glacial cycles, rather than the cause. Even so, they would tend to reinforce the temperature fluctuations, with higher concentrations of the greenhouse gases keeping the earth slightly warmer during the interglacial periods, and lower concentrations permitting further cooling during the glacial episodes.

It should be obvious even from this brief treatment that there are many possible answers to the question, What causes global glaciation? Because there are so many different parameters involved, each one interacting with the others, the development of large, high-speed computers has been a boon to research on the glacial-age climate. With them, it has been possible to simulate how climatic conditions should respond to different amounts of CO_2 in the atmosphere, different continental configurations, different parts of the Milankovitch cycles, and a host of other potentially important factors. The scientific literature is full of papers discussing General Circulation Models (GCMs, as they are known to the initiated), which can predict temperature distributions, wind patterns and many other climatic features for various possible past conditions. Much useful insight has been gained from the mathematical models developed for these simulations. However, like a long-range weather forecast, they are very sensitive to small changes in the input conditions, and their predictions are only accurate to the extent that the modelers have figured out in the first place how all of the parameters interact. Ultimately information from the earth itself, the record in the rocks reflecting the actual climate changes that occurred, is the standard against which these theoretical treatments must be judged.

GLACIAL CLIMATES, HUMAN EVOLUTION, AND THE RISE OF CIVILIZATION

The oldest known hominid fossils (*Hominidae* being the biological family to which our own genus, *Homo*, belongs) are very close to 4.4 million years old. They are found in Ethiopia, closely associated with volcanic ash deposits that can be dated quite accurately, so their age is well known. They are most likely our direct ancestors.

About 800,000 years after these early hominids lived, a remarkable fossil record of a different sort was being formed in what is now Tanzania, nearly 2,000 kilometers away from the site of the Ethiopian fossils. There, a series of volcanic eruptions covered the countryside with layer upon layer of fine volcanic ash. After rains the ash was like wet cement, and every creature that walked on it left tracks, creating a vivid record of the animal life that flourished in this part of Africa. But in addition to the tracks of everything from rabbits to elephants, there is something more in this snapshot of life from more than 3.5 million years ago: a group of hominid footprints wandering across the landscape. Most probably the creatures who left these tracks were similar to those represented by the older Ethiopian fossils. Some researchers who have studied these fossil tracks believe that they were made by a family group—mother, father, and child—but perhaps more importantly, the footprints indicate that these early hominids were walking on two feet, like modern humans. By about four million years ago, or even earlier, our ancestors had descended from the trees of the tropical forests of Africa and were spreading out onto the grassy plains, walking upright. Many paleontologists believe this transition was prompted by the gradually increasing aridity that occurred in Africa as the global climate cooled, reducing the extent of forests and increasing the area of grasslands. However, the true rigors of the Great Ice Age were still to come, although they may not have been as serious in the Tropics as in high latitudes.

The australopithecines, as the Ethiopian fossils mentioned above (and other similar hominids) are known, had small brains. They may have been bipedal, but they were probably not particularly smart. Nevertheless, they survived for several million years, for part of that time in parallel with our own genus. *Homo* first arrives on the scene among the hominid fossils in Africa about two million years ago. At roughly the same time, shaped stone tools appear in the sediments. One of the most distinctive characteristics of the new hominid is its large brain, large at least by comparison with those of any of the species of *Australopithecus* that preceded it. Why did *Homo* appear just at this time, and why was its brain larger than that of earlier hominids? As is the case with many other matters discussed in this book, there is no definitive, agreed-upon answer to this question, but there are many theories. One of these suggests that the correspondence between the appearance of *Homo* and the onset of Northern Hemisphere glaciation is no coincidence. In this view, the changing climate, particularly the alternation between long glacials and short interglacials, would have favored those with the ability to adapt to change, individuals with ingenuity and intelligence. In Africa the glacial episodes were cool and dry, and life would have been more difficult than during the warm, relatively wet interglacials. Whether or not this interpretation is correct is unknown. But the very great environmental changes that accompanied the cycles of cold and warm climates that have occurred with regularity over the past few million years must have played their part in forcing migration and isolation of population groups of both *Homo* and other animals. Rapid evolution of new species and subspecies, very apparent among the mammals as a whole, and certainly for the genus *Homo,* was an inevitable result.

By about a million years ago, a species of *Homo, Homo erectus,* had migrated out of Africa and into Europe and Asia. Fossils are sparse and anthropologists and paleontologists have had a difficult time trying to trace the lineage of modern humans, but it is known that by 100,000 years ago, early in the most recent glacial episode, a group of *Homo sapiens* known as the Neanderthals were living in Europe and the Middle East. In spite of the caricatures of Neanderthal as a dim-witted caveman with a club, these people had large brains—as big as our own—lived communal lives, and must have been quite intelligent. In Europe, they lived in a climate that was steadily deteriorating toward the maximum cold of the glacial period. However, Neanderthal disappears from the fossil record about 30,000 years ago, to be replaced by essentially modern humans, the Cro-Magnon people. These humans had arisen in Africa tens of thousands of years earlier, spread into Europe about 45,000 years ago, and for a while coexisted with the Neanderthals. In contrast to the Neanderthals, they apparently made sewn clothing and at least crude shelters, and were probably better equipped for the harsh climate. They experienced the cold conditions of Ice Age Europe firsthand, and also left beautiful cave paintings that give us an eyewitness impression of some of the now-extinct animals that roamed the glacial landscape, such as the huge, tusked woolly mammoth.

In addition to the climate itself, an important influence of glaciation on humans has been the fluctuations in sea level that have accompanied the glacial cycles. The low levels of the Wisconsin glacial maximum exposed very large areas of dry land that are now water-covered, in places providing migration routes for early humans, and for other animals as well. Australia and New Guinea were connected by dry land. Much of Indonesia was accessible on foot, or by very short journeys over water, and *Homo sapiens* migrated there from Asia. Perhaps the best-known result of the lowered sea level is the peopling of the Americas during the last glacial maximum. Twenty to thirty thousand years ago, it was possible to walk across the Bering Strait from Russia to Alaska. Mammoths and other large mammals migrated to North America via the Bering land bridge, and, near the peak of the Wisconsin glaciation, they were followed by curious Siberian tribesmen. Although much of eastern Siberia and Alaska were ice free, most of the rest of northern North America was covered in glacial ice, and the new immigrants were blocked from traveling to the east or south until the climate warmed into the present interglacial and the glaciers melted back. There is still controversy about the exact chronology of these migrations, but it is generally believed that as temperatures rose a north-south corridor opened up between the glaciers of the Rocky Mountains in the west and the ice sheets retreating toward Hudson Bay in the east, permitting migration to warmer, southern climates. We do know that by about 12,000 years ago there were people in the southwestern United States, and by 10,000 years ago they had spread into South America.

Although our immediate ancestors suffered through the hardships of the Wisconsin glacial period, human civilization as we know it has developed during an interglacial episode. Even so, the climate has not been as stable and equable as we tend to assume from the viewpoint of the short time span of our own lifetimes. With increasing detail, paleoclimatologists have constructed an impressive account of the climate over the past several thousand years using evidence that ranges from written historical records to variations in the thickness of growth rings in ancient trees. There is no doubt from these studies that there have been large fluctuations in both regional and local climatic conditions. The question that is the subject of fierce debate is the degree to which these changes have altered the course of civilization. The problem is the same one that faces those studying extinction events in the distant geologic past: linking cause and effect.

We know that even local, short-term climate variability causes great stress on human populations, a recent case in point being the dust bowl years of the 1930s in the central United States, when drought, coupled with poor agricultural practices, caused economic deprivation and eventually forced the migration of thousands of Oklahomans to California, an episode of American history immortalized in John Steinbeck's novel *Grapes of Wrath*. But much larger climate changes have affected the planet since the beginnings of civilization. Only a few of them can be touched on here.

The beginning of agriculture is generally taken to be a sign of the initiation of civilization. By this definition, civilization began in both the old and new world at about the same time. Some 6,000 to 7,000 years ago, there is evidence for domestication of sheep and the cultivation of crops in the Middle East. At approximately the same time, the people of southern Mexico began to raise corn. Climate studies show that this time was also the climatic optimum of the present interglacial, with global average temperatures significantly warmer than today, and also much more rainfall almost everywhere on the earth. In fact there is no evidence for *any* large deserts at this time. Is this just another coincidence, or is there a link between this benign climate and the emergence of civilization?

Several thousand years after this climatic optimum, around 4,200 years ago, a thriving civilization, the Akkadian empire, that had sprung up in the Middle East roughly between present-day Turkey and the Persian Gulf, suddenly collapsed. In its northern regions, agriculture diminished rapidly. According to records that have been found on clay tablets, much of the population migrated to the southern cities of the empire, along the Tigris and Euphrates Rivers, creating a refugee crisis that sorely taxed the government of the day. For decades, archaeologists have puzzled over the reasons for these events. But recent work shows that the beginning of this crisis coincides with evidence for a sudden drought, a drought that lasted some 300 years, in the northern parts of the Akkadian empire. Such a climate change would explain the well-documented migrations, because the agriculture-dependent northern regions relied on regular rainfall, and were without well-developed irrigation systems. In the south, the Tigris and Euphrates Rivers provided much more stable water supplies.

It is difficult to pinpoint a cause for the inferred climate change that apparently affected the Akkadian empire, and at any rate some archaeologists have argued that climate change alone would not be sufficient to explain the rapid demise of this civilization. But closer to the present we have even better-documented evidence for abrupt climate change and its influence on human life. A little more than 1,100 years ago, near the end of the ninth century, the climate in the North Atlantic region warmed, and stayed relatively mild for some 300 years or more. Climatologists have dubbed this period the "little optimum." In addition to the historical records (which only occasionally mention details of the climate), the year-by-year oxygen isotope variations in the Greenland ice cores confirm that a warm spell really did occur during this time. It was during this period that the Vikings, seafaring Norse adventurers, settled parts of Greenland. From there, in the relatively ice-free waters of the North Atlantic, they ventured west and reached North America. A well-preserved (and now restored) Viking settlement dating to about the year 1,000 exists in Newfoundland, which is probably the land referred to in Viking sagas as Vinland. The Vikings didn't stay long in North America, however; among other things they had to compete with the native Americans, who had arrived on the continent thousands of years before, not from Europe but from Siberia.

By late in the fourteenth century the climate in the North Atlantic region had again deteriorated to the point where there was at first infrequent, and then virtually no, contact between Scandinavia and the Viking settlements in Greenland. Eventually, those who still remained perished. The "little ice age" that followed the little optimum lasted from about 1450 to 1850, and had effects far beyond Greenland. During the climatic optimum, European agriculture and population had expanded, but in the following cold period there were floods, famine, and the plague. Especially in some northern regions where crops had flourished during the climatic optimum, the cold weather precipitated repeated harvest failures. Farms were abandoned, many rural areas deserted, and civil unrest occurred periodically. The famine-weakened populations could not easily fend off the plague. The historical record clearly documents the severity of the climate in Europe: The Dutch masters painted skaters on the canals of Holland, and in the seventeenth century there were frequent winter "frost fairs" on the ice of the River Thames in London. The river has not frozen again since 1814.

The climatic variations just described were of short duration, too short to be linked directly to the much longer glacial cycles. The record we have of them is also fairly local—it comes from Europe and the North Atlantic area. Most workers believe that they were caused by changes in the pattern of ocean circulation, in particular the amount of warm water flowing into the North Atlantic from the south. The causes of such sudden circulation changes, and whether or not they are characteristic of interglacial periods, is not known, although recent work on ice cores from Greenland suggest that the previous interglacial, which occurred about 130,000 years ago, exhibited even more short-duration climatic variability than the present one. Perhaps it is simply luck that has blessed us with a reasonably stable climate during the rise of industrial society over the past century and a half.

In a broad sense, modern humans are truly the progeny of the Great Ice Age. Our genus, *Homo*, appeared in Africa after Northern Hemisphere glaciation began, and the spread of our species to every continent on the globe occurred during the Wisconsin glacial period, with its lowered sea level. It is often difficult to untangle cause and effect, but as we have seen, variability of the interglacial climate of the past 10,000 years appears to have significantly influenced the course of human civilization. But 10,000 years is a very short time on the geologic scale. If there is one lesson to be learned from studying geologic history, it is that change is a constant on every timescale by which we can examine the earth: evolutionary change, change in the configuration of continents and oceans, change in the climate. In this book's short journey through geologic time we have examined just a very few of the changes that have occurred during our planet's four and a half billion years of existence. The geologic record, the record in the rocks, ends at the present, near the end of a warm interglacial period of the Great Ice Age. All that remains is to ask, What changes can reasonably be expected in the future?

13

WHAT COMES NEXT?
GEOLOGY AND MAN

IN THE VERY long term, the fate of our planet is clear. It will be engulfed in the searing fires of the sun as it expands and becomes a "red giant" star. Like all stars, the sun is fueled by nuclear reactions in its dense interior, where atoms of hydrogen are smashed together so tightly that they fuse to make heavier elements, with the release of huge amounts of energy. We know from observing other stars in the universe that when the hydrogen is used up in this process, the inner part of the sun will collapse into an even denser core, while an outer, "cooler" (but still at temperatures of thousands of degrees) envelope will expand into the solar system far beyond the orbit of the earth, consuming all in its path. But this will happen billions of years in the future, about as far from the present as the earth's creation is in the past, and just as difficult to imagine. By then, our species will be long gone.

There are some other certainties about the earth's future as well. The heat in the interior that drives plate tectonics, partly originating from radioactive decay and partly remnant from the earth's formation 4.5 billion years ago, is slowly diminishing, but at a rate so gradual that the geologic processes it fosters are likely to continue in much their present form for billions of years, perhaps until the dying days of our planet. Ocean basins will form and disappear, continents will collide, creating great mountain ranges that will then be ground down again to sea level by the onslaught of chemical and physical erosion, and, when conditions are right, the earth will again be plunged into ice ages. And as it travels through space, our planet will almost certainly collide head-on with pieces of the space debris that clutters our solar system. While not so very large on a cosmic scale, these fragments will be big enough for the impacts to change conditions on the earth's surface profoundly, if relatively briefly in geologic terms.

But on a shorter timescale, over just the next few generations of life on earth, there will be other, more immediate, pressures on our planet. One of my colleagues is fond of saying that the most important agent of geologic change at this particular instant in the earth's history is man. We are the first species with the capability to modify the planet's surface, its atmosphere, and its climate in a global and drastic way. Figure 13.1 shows how the human population of the earth has changed with time, and also how just one of our impacts on the environment—the addition of carbon dioxide to the atmosphere—has followed in step. In the past, much greater changes in carbon dioxide than are shown in this illustration have occurred due to natural causes. But as far as we can discern they took place considerably more slowly, and their effects, although severe for some plants and animals, were not imposed on a complex society like ours, which is very delicately adapted to the average climate of the last hundred years or so. If, as many scientists predict, the increasing atmospheric CO_2 causes a global temperature rise of a few degrees, the consequences will be quite staggering. Entire productive agricultural belts will disappear from use, or at the very least will be suitable only for quite different crops than are presently raised. (On the other hand, high latitude regions now of marginal use for agriculture, especially in Russia and Canada, may suddenly have great potential for food production.) Sea level will rise with the increasing temperature, partly because of further melting of the ice caps and partly because the ocean water itself expands as it warms, flooding many highly populated, low-lying regions and increasing the vulnerability of others to tropical storms. As long as man is wedded to the use of fossil fuels, the burning of which is the main source of the CO_2 added to the atmosphere, there is little possibility of stemming its increase, although with concerted international action it may be

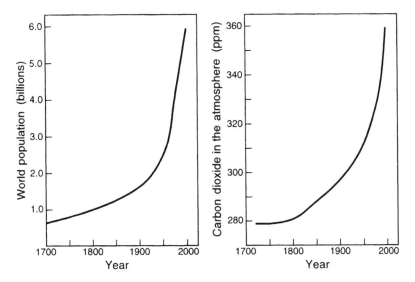

FIGURE 13.1 *Graphs showing the changes in world population (left) and carbon dioxide content of the atmosphere (right) since about 1700, using data from various sources. The CO₂ concentrations are expressed in parts per million (ppm); by the mid-1990s, carbon dioxide constituted about 360 parts in a million of the atmosphere. Although the* **percentage** *changes are vastly different in the two cases, it is clear that the rate of increase of both population and atmospheric carbon dioxide has risen rapidly in the second half of the twentieth century.*

possible to slow the rise somewhat. Over the long term society will undoubtedly adapt to the changes that are bound to occur. However, because these changes will be rapid even on a human timescale, it is likely that they will cause severe hardship and disruption in many parts of the world.

There is a possibility that the increase in temperatures that will certainly result from added atmospheric carbon dioxide will counteract the tendency for the earth to plunge into another glacial era, which, based on long-term records such as that in Figure 12.4, is just around the corner. However, there is little likelihood that the two effects will exactly balance one another. Most scientists who have studied the question believe that the CO_2-induced warming will win out, and that we are in for a "superinterglacial" period that will last until most of our fossil fuel resources are consumed. By that time, several centuries into the future, the concentration of carbon dioxide in the atmosphere will be at least three times its value in preindustrial times. Gradually, much of this extra CO_2 will be absorbed by the oceans, and in the absence of new inputs its atmospheric concentration will decrease, eventually allowing the earth to slip into a slightly delayed glacial period.

In the long-term, geologic, view of the earth, changes wrought by man, such as increased atmospheric CO_2, are minor perturbations. As should be obvious from the preceding chapters, the earth has undergone much more severe environmental disturbances in the past, but even so their legacy in the rock record is generally quite subtle. Were humans to disappear from the planet tomorrow, the signs of our occupation some millions of years hence would be quite minor. But our limited individual lifespans tend to focus most people's attention on the shorter term, and on such timescales, because we have a good understanding of how the earth works, it is possible to make some predictions about what may be in store.

OUR FINITE GEOLOGIC RESOURCES

The development of geology as a science was largely founded on the search for raw materials. Until quite recently, the expectation of most people trained professionally in the earth sciences was that they would work in the petroleum or mining industries. Indeed, enrollment in geology departments at colleges and universities across North America very closely followed the ups and downs of the major oil companies, who were the main employers of graduates. However, after a very long period of depressed prices for petroleum, and with increased emphasis on conservation and the environment in recent years, that picture is changing. But it is still the case that finding and extracting the materials that are necessary for our complex world are important aspects of the earth sciences. It is also an area in which prospects for the future are fairly certain.

From ancient times, prospectors (for want of a better word) have used intuition, experience, and brainpower to locate geologic materials that are useful and in demand. In modern times, technology, especially remote sensing technology, has been added to their arsenal, allowing the search for resources to extend to remote areas not previously easily accessible, and to the ocean depths and the subsurface regions of the continents. Why are such efforts necessary in the search for these resources? The answer is that while there are small amounts of almost every chemical element in the periodic table in even the most ordinary of materials—there is gold dissolved in the ocean, copper present in the soil in your garden—they are there in very dispersed form, not amenable to recovery in ways that are economically feasible. Even aluminum, the third most abundant element in the earth's crust, cannot be mined just anywhere. Over the earth's history, however, geologic processes have acted to concentrate aluminum and many other elements, forming valuable deposits. The trick has been to discover how these processes work, and to use that knowledge to narrow the search for deposits that can be mined at a reasonable cost. Exploration for such finds, ever more sophisticated, is still going on, but large tracts of the earth have already been searched in considerable detail and the rate of new discoveries has slowed. New technologies permit the extraction of desired materials from deposits that were once thought worthless, but even so, geologic resources are not inexhaustible. Their concentration has occurred over several billion years of earth history, and on a human timescale they are not renewable. In some cases, we are depleting them over decades.

Perhaps the most stunning example of this phenomenon concerns petroleum. Because of its extreme importance for modern society, the formation and distribution of petroleum deposits has been studied in great detail, and billions of dollars have been spent on exploration for and recovery of this natural substance. Although crude oil occurring as surface "seeps" had been known for thousands of years, and used for a variety of purposes, the very first oil well was drilled in Pennsylvania in 1859. Known at the time as Drake's Folly, this small venture spawned a gigantic, global industry that has touched virtually every corner of the earth. But within a century of this first well, with millions of barrels of oil being pumped from the earth every day, a few cautious voices began predicting that dire consequences would follow our unfettered consumption of this nonrenewable resource. Although some of the more extreme predictions have not come to pass, mainly because of more efficient energy use, a global economic slowdown, and the discovery of new deposits, there is no question that we will eventually run out of oil and natural gas. The only uncertainty is how long it will take. Although formed over millions of years, geologic resources of petroleum will be available for abundant consumption for only a few *hundred* years!

The resource that we use so lavishly, the substance that fuels our automobiles, is actually ancient solar energy, stored by nature as petroleum. Chemically, it is mostly carbon, combined with 15 to 20 percent hydrogen. It forms only in very specific geologic environments, namely in the muddy sediments deposited in warm, shallow seas. In such places the organic remains of plankton, the small, floating organisms that live in the sunlit surface ocean, accumulate rapidly on the seafloor and are buried. Burial protects the organic remains from destruction, but the processes that turn this dispersed, carbon-rich material into petroleum are complex. The key factors seem to be temperature and time. As the organic matter is buried progressively deeper, the temperatures to which it is subjected increase. It appears that the most favorable temperature range for the production of liquid petroleum lies between about 65°C and 150°C, typically corresponding to depths of several kilometers. But even if the organic material is transformed into oil, it is not easily extractable from the fine-grained sediments in which it formed. Only when it is found in coarse-grained materials with abundant pore space, such as sandstone, can it be readily recovered. Fortunately, oil is a low-density liquid—it floats on water—and with time it migrates upward, sometimes into adjacent rock formations. Most productive oil fields are thus found not in the rocks where the petroleum originated, but in nearby, porous strata.

Even such rudimentary information about oil formation simplifies the task of petroleum exploration enormously. Because abundant marine life is required for its production, rocks dating from before the Cambrian period, when life was sparse, are unlikely to hold oil. The same is true for strongly metamorphosed rocks of any age, because they have experienced temperatures high enough to destroy any petroleum they may once have held. The primary targets for oil exploration have thus been thick accumulations of Phanerozoic sediments formed in shallow seas along the margins of present or former continents, or in the inland seas that periodically flooded parts of the continents. The accumulated experience from drilling in such regions also makes it possible to predict reasonably accurately just how much oil and gas is still undiscovered. It is these predictions, coupled with estimates of how rapidly petroleum resources will be consumed in the future, that suggest that a century from now mankind will have used up most of the earth's petroleum (see Figure 13.2). Unfortunately, the relatively rosy short-term prospects for petroleum supply compared to demand tend to obscure the need for concern about long-term shortages, which will surely occur. Other sources of energy—and of raw materials to replace the petroleum used for products as diverse as polyester clothing, fertilizers, and drugs—will have to be developed. It would be better if this happened sooner rather than later.

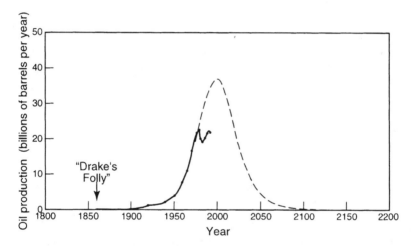

FIGURE 13.2   *The solid line shows a smoothed curve of the world's oil production, from the time the first well was drilled in 1859 until 1991. Dots along the curve are actual production figures from the* annual International Petroleum Encyclopedia *(PennWell Publishing Company). The dashed line shows an "optimistic" prediction of future production that was made in 1969 by U.S. Geological Survey geologist M. King Hubbert, based on his best estimate of the amount of known and undiscovered oil reserves. Although the downturn in petroleum usage of the early 1980s will push the time over which oil is abundant a bit further into the future, it is still clear from this diagram that we are one of only a few generations of humans who will enjoy the benefits of this resource.*

Although oil and gas are startling and sobering examples, many other geologic resources are also being consumed at rates that are unsustainable. Furthermore, like oil and gas, the distribution of these materials depends on geologic factors, not political boundaries, making the dependence of modern industrialized societies on specific resources especially precarious. A good example is the element cobalt, a critical ingredient for the alloys used to make permanent magnets, jet engines, and other high-performance machinery. The United States, and indeed most developed nations, possesses essentially no native supplies of cobalt. In the late 1970s, civil war in Zaire, a major supplier country, pushed up the price of this commodity to more than ten times its previous value. The shortage was not long-lived, but it was nevertheless a sharp reminder of the finite nature of mineral resources.

Some space enthusiasts have suggested that the moon, or even the asteroids, might be future sources of raw materials for the earth. All of the necessary chemical elements are present on the moon, it is true, but their extraction would require enormous amounts of energy. In contrast to the situation on the earth, geologic processes on the moon have not, for the most part, acted to form mineral deposits as we know them. The reason is that a surprisingly large fraction of the concentration mechanisms on our planet depend upon the presence of liquid water. Some deposits precipitate directly from the sea—for example, the banded iron formations discussed in Chapter 4 that are the source of much of our iron ore. Others are the products of water-assisted weathering—aluminum is concentrated when the high rainfall and warm temperatures of tropical regions dissolve away nearly everything else in the local bedrock, leaving behind only the insoluble, aluminum-rich mineral bauxite. Gold and many other valuable metals commonly occur in veins because they were deposited there from hot, water-rich fluids flowing through cracks in the rocks of the earth's crust. For reasons that have to do with its origin, the moon is devoid of water, and most of the processes that concentrate minerals on the earth have never occurred there. As a result, economically valuable elements in the rocks of the moon are present only in dispersed form. At least in the near future, it appears that the voracious appetite of modern societies for raw materials will have to be met from earthly sources, by developing methods for efficient extraction of materials from less concentrated ores, by conservation and recycling of valuable materials, and by developing substitutes for some of the rarest minerals and elements. For this reason, the role of geoscientists has changed (at least in part) from one of being simply exploiters of the earth's abundant mineral deposits, to being stewards of the resources that are now recognized to be limited. Geologists have the necessary knowledge to assess the long-term consequences of consuming critical materials at current rates, and some have also assumed the responsibility of alerting both governments and the population at large to the probable results of such consumption.

THE IMPACT HAZARD

The geologic record leaves little doubt that there will be impacts in the earth's future. The current debate among scientists focuses on just what the probability of a very large, widely damaging impact is, and whether there are the means to avoid such a disaster.

The evidence for a gigantic impact 66 million years ago, creating a global crisis that may have led to the extermination of the dinosaurs and many other creatures, was discussed in Chapter 10. The body that caused the K-T disaster was probably an asteroid, perturbed into an orbit that happened to intersect that of the earth. It is generally recognized that very large events like the K-T impact are rare, even on geologic timescales, but what is often not appreciated is that there are hundreds, perhaps even thousands, of asteroids with diameters greater than about 100 meters (and therefore capable of doing substantial damage if they were to strike our planet) that at this very moment have orbits that cross the earth's. Each of these has some potential for striking the earth, and the geologic—and even the historical—record shows that such collisions have occurred regularly in the past. Just what is the probability that they will happen again, and how much damage will they do? Considerable effort is now being directed toward answering these questions. While there is always some uncertainty in predictions, the threat is real enough that there has already been serious discussion about the feasibility of early identification and possibly even diversion of objects that seem to be on a collision course with the earth. One recent analysis by Clark Chapman of the

Planetary Science Institute in Tucson, Arizona, and David Morrison of NASA's Ames Research Center in California, published in the scientific journal *Nature* in 1994, predicted that there is one chance in 10,000 that an asteroid large enough to disrupt the environment and kill a large fraction of the world's population will strike the earth during the next century. That is a very low probability, but in purely statistical terms, because of the very large number of deaths that would ensue, it indicates that for the average American the chances of dying in an impact are about the same as those of being killed in an airline crash. Air safety is a legitimate concern of both governments and citizenry; should asteroid detection be also?

The evidence necessary to make estimates of impact probabilities comes from a variety of sources, including the geologic record. Because the atmosphere shields us from small bodies, which burn up from frictional heating before they reach the surface, and because weathering and plate tectonics continually modify the landscape, our planet's surface is not densely pockmarked with craters as are some of our neighbors'. Nevertheless, there are many well-documented examples. Meteorite Crater, in Arizona, has already been mentioned; it is a relatively young and well-known impact structure. Many terrestrial craters are quite large, and were initially recognized from their circular shape only after being seen by airplane-borne or satellite observers. Ancient, worn-down parts of the earth's crust, such as the Precambrian Canadian Shield in North America, contain many old craters. Fortunately, burial beneath a covering of sedimentary rocks and soil helped to preserve them until the recent Northern Hemisphere glaciation scraped away the protective blanket, again exposing the craters at the surface. Careful examination of the sizes and ages of these and other craters has allowed geologists to build up a database documenting the frequency of impact for bodies of different sizes. Similar but even more comprehensive evidence has come from studies of craters on the moon (recall Figure 3.1), which has no protective atmosphere to destroy smaller bodies, and where the processes that destroy craters on the earth, such as weathering or plate tectonics, do not operate. Thus large parts of the moon's surface have acted as inert recorders of all impacts over billions of years.

The accumulated evidence indicates that at the very small end of the size spectrum, a body with about the energy of the nuclear bomb that was dropped on Hiroshima at the end of the Second World War collides with the earth *every year!* In terms of actual physical size, these objects are small; the atmosphere protects us from their effects and they burn up or explode very high above the surface due to friction. Except for the fact that they are recorded by surveillance satellites, we are not even aware of their existence. Even objects with a hundred times this energy, one or two of which can be expected every century, do not reach the earth's surface. But there are much bigger asteroid fragments floating around in the vicinity of the earth, and in the relatively recent past there have been well-documented cases both of near misses, and of actual strikes. The near miss occurred in 1989 when an asteroid several hundred meters in diameter, estimated to be carrying the energy equivalent of more than 1,000 megatons of TNT, passed us at less than twice the distance to the moon. A slightly different orbit could have sent this object crashing into the earth, with disastrous results. It would certainly have reached the surface, creating a crater many kilometers in diameter (or generating gigantic waves had it landed in the ocean). The near-miss asteroid, had it actually collided with the earth, would have been about a hundred times more destructive than the object mentioned in Chapter 3 that exploded in the atmosphere over Siberia in 1908. The Tunguska event, as it is known, was recorded in Europe from atmospheric shock waves, and when scientists eventually reached the remote site many years after the fact, they found forests flattened over an area of more than 2,000 square kilometers, and evidence that the explosion had touched off fires near the center of the affected region. No fragments of the asteroid have ever been found, but recent calculations suggest that it was probably a stony object that exploded in the atmosphere at an altitude of about 10 kilometers. Fortunately, the countryside where it struck was uninhabited; had it fallen in a populated area, the consequences would have been severe.

Large impacts are rare, but their effects are potentially so devastating that they constitute an entirely different type of geologic hazard than almost any other. Collisions like the one that ended the Cretaceous period throw up so much fine debris into the atmosphere that—except for the light of the fires ignited by the impact—the entire globe would be plunged into pitch blackness for some period of time. Even much smaller impacts could still reduce the sun's illumination to the point that agriculture might be shut down for more than a growing season, with disastrous results. Societies globally would be affected; no countries would be spared and capable of helping others in need, as they do for floods, earthquakes, or drought. Fortunately, the really large objects, with diameters of a kilometer or more, are the ones that are most easily detected in space. The technological capability exists to find and track the orbits of such bodies. Fortunately also, sensitive telescopes and thorough surveys could probably provide enough advance warning (several years, at least) that it would be possible to design and carry out a defensive strategy to avoid impact from those objects found to be on a collision path with the earth. Diverting an asteroid would be a very expensive proposition, but it is unlikely that there would be many complaints about the cost of protecting the earth from potential devastation.

VOLCANOES AND EARTHQUAKES

Much more immediate, if more localized, danger to society exists from large earthquakes and volcanic eruptions. These are the phenomena most people think of when they imagine geologic disasters. With present knowledge about how the earth works it is a relatively straightforward matter to make predictions about the likelihood of such events. It is possible to say with almost 100 percent certainty that sometime in the next few hundred years a large and very damaging earthquake will strike San Francisco, or Tokyo, or that Mount Saint Helens will erupt. But it is not yet possible to predict, long in advance, precisely when any of these events will occur, or, equally important, just how large they will be. Progress is being made on shorter-term prediction, however. In most cases this involves careful monitoring, using both instruments and simple observation, in regions that are already known to be at high risk. In a few instances, large-scale evacuations have been carried out when the danger was believed to be immediate.

Perhaps the best-known example is the 1975 evacuation from the volcanic island of Guadeloupe in the Caribbean, where ominous precursors suggested that an eruption was imminent. However, it wasn't. Three months later, the inhabitants were returned to their homes, no disaster having occurred, and heated debate raged about the wisdom of the evacuation, and, of course, the accuracy of the predictions. But nature is capricious, and it will be a long while before we understand with certainty exactly which kinds of signals truly presage an eruption or an earthquake. In the meantime, there are likely to be other mistaken warnings, but in the long run it is probably better to heed than to ignore them. This was brought home with a vengeance not long after the Guadeloupe episode, when geologists in Colombia warned that even a small eruption of the volcano Nevado del Ruiz could melt the snow and ice at its summit, triggering great flows of volcanic ash and mud, which would endanger the town of Amero at its base. In this case the warnings were ignored, and the predicted mudflows occurred only months later, killing 25,000 people.

As should be clear from the discussion of plate tectonics in Chapter 5, the potential for both volcanic eruptions and earthquakes is highest along the plate boundaries. Places where the plates converge at subduction zones are most at risk. A quick glance at Figure 5.2 will also show that many such regions are heavily populated: Much of the west coast of North, Central, and South America, Japan and Indonesia, and parts of the Mediterranean lie close to subduction zones. All of these areas have experienced earthquakes and volcanism in recorded history, and will again in the future. In most, however, the disasters occur at rather large intervals, often of a generation or more apart, and therefore are not very prominent in the general consciousness.

Even when relatively near-term geologic danger is quite certain, the response may at best be muted. San Francisco, one of the most beautiful but also one of the most deadly cities in the United States in terms of earthquake hazard, is also perceived to be one of the most desirable places in the country to live, and has correspondingly exorbitant real estate prices. Although the city does not lie at a subduction zone, the San Andreas Fault passes directly through it, and several other large faults exist in the region. The infamous 1906 earthquake (caused by a rupture along the San Andreas itself) and the ensuing fires, which together destroyed most of the downtown area, are still referred to frequently in the press, but the implications are sloughed off by most present-day residents, who prefer to enjoy the beauty of the city and gamble that the next "big one" will not strike in the immediate future. But inevitably, driven by plate tectonics, it will occur, and although modern building codes ensure that damage will be less than it would be in their absence, they are no guarantee of safety. A 1989 earthquake much smaller than that of 1906 and occurring nearly 100 kilometers south of the city, near Santa Cruz, California, damaged buildings and bridges in San Francisco and its vicinity, and resulted in sixty-five deaths. Many other major world cities face similar risks from geologic processes, their occurrences a virtual certainty on timescales ranging from a few tens to a few hundreds of years.

Fortunately, earthquake destruction is quite localized. However, when they occur in the sea, earthquakes can generate large tsunamis, which may travel across entire ocean basins and cause extensive damage in far-flung parts of the globe. Although these gigantic waves travel quickly, there is usually sufficient warning that residents of coastlines likely to be affected can prepare, at least to the extent of moving away from low-lying areas. Very large volcanic eruptions can also have consequences far beyond their immediate vicinity. It was noted in Chapter 12 that the 1991 eruption of Mount Pinatubo in the Philippines caused globally lowered temperatures for several years due to the volcanic aerosols, mainly sulfur dioxide, that were injected into the atmosphere. Immediately after the initial eruptions there was so much volcanic debris in the atmosphere that commercial airliners plying routes across the Pacific reportedly had to have their windshields replaced every few days because of pitting. The same dust was responsible for spectacular sunsets around the world for more than a year.

Many eruptions in the past have left easily traceable ash layers in the geologic record, often several centimeters thick and spread over tens of thousands of square kilometers. The largest in recent history occurred in 1815, on the island of Sumbawa in Indonesia, when the great volcano that was Mount Tambora erupted violently. According to the records of European officials in the region at the time, explosions accompanying the eruption were heard 1,500 kilometers away. In Java, hundreds of kilometers to the west of Tambora, the daytime was as black as night because of ash. The volcanic debris thrown into the atmosphere was also almost certainly responsible for unusually cold weather throughout the globe that followed the eruption. In a delightful little book on the subject of volcanoes and climate, Henry and Elizabeth Stommel carefully documented the cold, wet (and even snowy) summer of 1816 in New England, Europe, and elsewhere that followed the Tambora eruption. In their research they frequently ran across a colloquial expression of the time that referred to that frigid summer: "Eighteen hundred and froze to death."

There are enough data from recent, carefully monitored eruptions such as Pinatubo that it is clear that the huge amount of ash and sulfur dioxide from Mount Tambora would have had a significant effect on the amount of solar energy reaching the earth's surface, resulting in substantial cooling of the earth. Indeed, some researchers have noted that the largest volcanic events evident in the geologic record, which are many times the size of the Mount Tambora eruption, would have been capable of producing "volcanic winter," perhaps lasting several years. It is virtually certain that such eruptions will occur in the future, and they could be almost as destructive to society as the large impacts discussed in the previous section. It is also possible that global cooling following such events, if it occurred when conditions were otherwise appropriate for glaciation, might provide the push needed to tip the earth into an ice age.

Geology, it is clear, does not respect national boundaries. Instead, its bounties in the form of mineral and energy resources from the earth, as well as its hazards, are the present-day manifestations of geologic processes that have been going on for millions if not billions of years. These processes can alter the face of the earth quite radically, change the climate in dramatic ways, and even influence the course of evolution. We know all of these things from studying the geologic record, the record preserved in the rocks. As that record is unravelled in increasing detail, it should be possible to predict what lies ahead with ever greater certainty. It should also be possible to determine how the actions of that most recent agent of geologic change to appear on the scene, man, are likely to perturb the ongoing natural geologic cycles. And perhaps equally satisfying, it will permit us to perceive with even more clarity the origins of the landscapes, brimming with geologic history, that surround us every day of our lives.

GLOSSARY

accretion   With reference to the early earth, the process by which solid material orbiting the sun aggregated to form the earth. The individual fragments of material may have ranged from sand-sized grains to planetlike objects the size of Mars.

andesite   A type of volcanic rock that is characteristic of volcanoes formed at subduction zones. The name comes from the Andes mountains.

angiosperms   The flowering plants. The great evolutionary advantage of these plants is that the flowers attract insects, which carry pollen from one flower to another.

Archaeopteryx   An animal, now extinct, that had characteristics both of birds and reptiles. It is considered to be one of the first true birds because of its feathers and bone structure. *Archaeopteryx* lived fairly late in the Jurassic period.

asteroid   Small (up to almost 1,000 kilometers in diameter for the largest known asteroid), rocky, and metallic bodies that orbit the sun. They are concentrated in the asteroid belt between Mars and Jupiter. Most meteorites are believed to come from the asteroid belt.

basalt   A very common fine-grained, dark-colored igneous rock that forms by cooling of erupted volcanic lava. The lavas originate by melting within the mantle. Basalt is the predominant rock type of the seafloor and of many oceanic islands, such as Hawaii, and is also common on the continents.

bauxite A type of rock produced by weathering in tropical climates that is a major ore of aluminum. It is composed almost entirely of hydrated oxides of aluminum, virtually all of the other components of the original rock having been leached away by abundant, warm rainfall.

chert A hard, very fine-grained sedimentary rock composed primarily of silica (SiO_2). It is also known as flint. Chert may be formed by direct precipitation from water, or by accumulation of the silica-rich skeletons of some plankton.

chondrite A common type of meteorite, believed to be composed of bits and pieces of the earliest-formed materials in the solar system. Chondrites apparently come from small bodies that never underwent melting and chemical differentiation, and therefore are valuable clues to the nature of the original material that accreted to form the earth.

conglomerate A sedimentary rock composed mostly of rounded pebbles and boulders, often of a variety of rock types, set in a finer-grained matrix. It is essentially the lithified equivalent of a stream gravel.

coral The general name for a large group of marine invertebrate organisms that live in shallow water and make their skeletons from calcium carbonate. Corals are frequently colonial, making coral reefs, and they are common in the fossil record.

core The innermost part of the earth, lying beneath the mantle (see Figure 1.2). It is composed primarily of an iron-nickel alloy, but also contains some lighter elements. It has an inner, solid portion and an outer, liquid one. The earth's magnetic field is thought to originate in the outer portion.

cyanobacteria A type of bacteria found fossilized in Archean rocks, but also with living examples (present-day stromatolites are composed of cyanobacteria). Like all bacteria, the cyanobacteria consist of primitive (prokaryotic) cells with no internal structures.

detrital An adjective used to describe fragments of rock and mineral material that have been eroded from a parent rock and transported to a site of sediment deposition.

erratic The term given to a rock or boulder that has been transported far from its origin by glaciers. In general, erratics are distinguished because they are a different rock type than the local rocks.

eukaryote An organism with cells containing a nucleus, chromosomes, and other internal structures (eukaryotic cells). This type of cell typifies all organisms except the bacteria and the cyanophytes.

evaporite A chemically deposited sedimentary rock, typically composed mostly of ordinary salt (NaCl) and gypsum (CaSO$_4$), that has been formed by direct precipitation from an evaporating body of water.

fault The surface along which rocks have broken, and the two portions moved relative to one another. Essentially a crack in the earth's crust along which movement has taken place.

genus The taxonomic category above *species*. Similar species belong to the same genus.

granite A common, coarse-grained igneous rock of the continental crust, typically made up primarily of feldspar, quartz, and mica. Granite forms from magma that is not erupted at the surface but rather cools slowly within the crust. The slow cooling accounts for its coarse grain size.

greenhouse effect An increase in temperature at the earth's surface due to the trapping of heat by various constituents of the atmosphere, such as carbon dioxide and methane. Compounds that prevent the reradiation of heat energy from the earth have been dubbed greenhouse gases. The obvious analogy is to real greenhouses, in which it is clear glass that traps the outgoing radiation.

half-life The term used to describe the length of time it takes for one-half of the atoms in a sample of radioactive material to decay. The half-life is directly related to the decay constant, which is the probability of decay per unit of time.

hematite   A common and widespread iron mineral with the chemical formula Fe_2O_3. In some forms it has a distinctive red or red-brown color.

hot spot   The term used to describe the surface expression of a mantle plume. Hot spots are regions of extensive volcanic activity, high heat flow, and are generally elevated relative to their surroundings.

isotope   All isotopes of a chemical element have the same chemical behavior, but they differ from one another in the number of neutrons they have in their nuclei, and therefore they differ in atomic weight.

kimberlite   A peculiar type of volcanic rock, quite rare but also quite important because all known diamond occurrences originate in kimberlites. Because many kimberlites contain diamonds, they must originate by melting at great depths in the mantle, probably approximately 200 kilometers deep.

kinetic energy   The energy associated with any moving body because of its motion.

limestone   A sedimentary rock composed primarily of calcium carbonate. Some limestones are simple chemical precipitates, but in most the calcium carbonate was originally produced as the shells or skeletons of marine organisms, which accumulated on the seafloor when the organisms died.

lithified   Turned to solid, coherent rock.

lithosphere   The outer, rigid shell of the earth that makes up the plates of plate tectonics. On average, it is about 100 kilometers thick, and it contains both the crust and the uppermost part of the mantle. The boundary between the lithosphere and the underlying mantle is a mechanical boundary, not a chemical one, and occurs where the upper mantle is hot and quite plastic.

loess   A fine-grained, wind-blown sediment that occurs extensively throughout the world, especially in the Northern Hemisphere. Its origin is generally believed to be tied to the ice age of the past few million years, perhaps because of the predominantly arid climate and stronger atmospheric wind systems at the height of the glacial periods.

magma Molten rock formed by melting within the earth. It forms igneous rocks when it cools; if it is erupted on the surface it is termed *lava*.

mantle plume A jet of material with lower density than its surroundings, rising from deep within the mantle. Mantle plumes are probably also hotter than their surroundings, and as they approach the surface undergo melting, leading in many cases to surface volcanism. They are believed to be stationary and long-lived, producing volcanoes such as those of Hawaii and the long chain of now-extinct and submerged volcanoes that trail off to the northwest of Hawaii.

mantle That part of the earth between the crust and the core (see Figure 1.2). It makes up the bulk of the earth and is composed of dense minerals, mostly silicates and oxides of magnesium and iron.

mare The name given to the dark, low-lying regions of the moon. It is now known that the reason for the dark color of the maria is that they are floored by dark-colored basalt flows.

mass extinction An episode when the extinction rate of plant and animal species is especially high, usually lasting no more than a few million years. Most of the major boundaries in the geologic timescale coincide with mass extinctions.

metamorphism The process by which the mineral makeup of a rock is changed when it is subjected to high temperatures and pressures.

meteorite Rocky or metallic objects that fall to earth from space. Most are believed to originate in the asteroid belt.

microbe Literally, a small (microscopic) form of life. Used mostly in reference to bacteria.

monotremes A group of primitive, egg-laying mammals, most of which are now extinct. The platypus of Australia and two species of anteaters from Australia and New Guinea are the only living monotremes.

moraine A ridge or hill of glacial till, usually deposited at the margin of a glacier.

nappes   Huge, flat-lying folds in rock that are characteristic of the collision zones between continents. They are produced by the extensive thrusting that occurs in such zones.

outcrop   The term that geologists use for an exposed portion of a rock body or formation, visible at the earth's surface.

paleontology   The study of life in the past, largely through investigation of fossils.

passive margin   The edge of a continent that is entirely within a tectonic plate, and therefore not subject to the volcanism, seismicity, and other geologic processes that occur at plate boundaries. Typically sediments accumulate quietly along passive margins, as they do today along the eastern coast of North America, or the western coasts of Africa and Europe.

photosynthesis   The process by which plants convert carbon dioxide and water into sugars, with the release of oxygen.

phylum   A broad, primary taxonomic category of living things. There are five recognized kingdoms into which all living things can be placed; each of the kingdoms is divided into a number of phyla.

plankton   A general term for the small floating plants and animals of lakes and the oceans.

plate   In terms of plate tectonics, an approximately rigid segment of the lithosphere that moves about the earth's surface relative to other plates.

prokaryote   An organism consisting of a primitive cell (prokaryotic cell) without internal structures such as chromosomes or a nucleus. All bacteria are prokaryotes.

pterosaur   An extinct flying reptile that lived during the Mesozoic era.

pyrite   A mineral composed of iron and sulfur with the chemical formula FeS_2. It is widespread in the earth's crust, and is colloquially referred to as fool's gold.

quartz   A common mineral of the earth's crust, with the chemical composition SiO_2.

radiation In the evolutionary sense, radiation is expansion of a group of organisms into new habitats and environments, with an accompanying divergence of evolutionary traits.

sandstone A sedimentary rock usually composed predominantly of grains of quartz, cemented together by a chemically precipitated matrix such as hematite, calcium carbonate, or silica.

seismic Refers to earthquakes, or the vibrations they cause in the earth. For example, seismic wave is the general term used to describe the vibrations that travel outward from the point where an earthquake occurs; a seismic gap along a fault would be a section where no earthquakes occur.

shale A type of sedimentary rock composed mostly of fine-grained clay minerals. It tends to split easily along the original sedimentary bedding planes, forming thin, flat fragments.

species The taxonomic category that ranks below *genus* and is not subdivided. Individuals of the same species are capable of interbreeding and producing similar offspring.

stromatolite A bulbous, layered structure found in sediments. Stromatolites are especially prominent as fossils in Proterozoic sediments, but living varieties also exist today. The stromatolite structure is built up by colonies of filament-like algae that form layers of sticky threads, trapping sediment particles between them.

subduction The process by which a portion of a lithospheric plate descends into the earth's interior at a collision boundary between plates. The descending plate invariably is one that carries dense oceanic crust, and the subduction zone is marked at the surface by a deep oceanic trench. Subduction zones are characterized by large earthquakes.

supernova An exploding star. Supernovae occur when the "nuclear fuel" of a large star becomes depleted and the central part of the star collapses, leading to a cataclysmic explosion with the release of huge amounts of energy.

suture The region along which two fragments of continental crust have been joined together through collision. Initially such zones are marked by high mountains; the Himalayas and the Alps are relatively young examples of suture zones.

talus Also referred to as talus slope, or scree slope, it refers to an accumulation of loose debris forming a steep apron of rock fragments weathered from a cliff or steep mountainside.

till A general term that describes the unconsolidated debris deposited by glaciers. Till is usually a mixture of fragments of many different sizes, from boulders to microscopic clay particles, and also frequently of many different rock types. Till that becomes cemented into solid rock is referred to as *tillite*.

trilobite An extinct class of invertebrate animals that lived in the seas from the Cambrian period until the end of the Permian. The trilobites are related to crustaceans, insects, and spiders.

uniformitarianism The idea that geological processes that can be observed in the present are likely to be the same as those that operated in the past.

uraninite A uranium mineral with the chemical formula UO_2.

varve A layer of sediment deposited over the course of a year in a glacial lake. In typical cases there is a gradation from coarse material deposited from the summer runoff, to finer, usually dark-colored, material deposited by slow settling of the fine, organic-material-rich suspended sediment during the winter.

zircon A widespread but trace mineral of the earth's crust that has the chemical formula $ZrSiO_4$. Because it typically contains uranium, it is used extensively for uranium-lead dating.

FURTHER READING

For those interested in delving deeper into some of the topics discussed in this book, the following could serve as starting points. This short list is very far from being comprehensive, but particularly the recent textbooks mentioned below contain quite extensive reference lists of books and journal articles that deal with a very wide range of topics in the earth sciences.

Physical Geology

Physical Geology is geology without the historical context. It concentrates on the processes, physical and chemical, that shape our planet. Introductory courses in geology are often physical geology courses, and there are many good textbooks. A few recent editions with which I am familiar are listed below. There are several others, quite likely just as good, that I do not know as well.

The Dynamic Earth (3rd Edition). B. J. Skinner and S. C. Porter, John Wiley and Sons, Inc., 1995.

Earth, An Introduction to Geologic Change. S. Judson and S. M. Richardson, Prentice Hall, Inc., 1995.

Understanding Earth. F. Press and R. Siever, W. H. Freeman & Company, 1994.

Earth's Dynamic Systems (6th Edition). W. K. Hamblin, Macmillan Publishing Company, 1992.

Geologic History

The books listed below are textbooks that cover much of the material discussed in the present book.

Evolution of the Earth (5th Edition). R. H. Dott, Jr., and D. R. Prothero, McGraw-Hill, Inc., 1994. A well-illustrated and comprehensive textbook that covers all of historical geology.

Earth and Life through Time (2nd Edition). S. M. Stanley, W. H. Freeman and Company, 1989. Numerous photographs and drawings aid the development of the ideas presented in the text.

Life and Evolution

History of Life. R. Cowen, Blackwell Scientific Publications, 1990. A thorough, up-to-date, and well-written account of evolution, from the origin of life on earth to *Homo sapiens.* Written as an introductory textbook on the subject.

Wonderful Life. S. J. Gould, W. W. Norton & Company, 1979. A discussion of the nature of evolution from the point of view of the Cambrian Explosion, specifically the fossils of the Burgess Shale of British Columbia.

Extinction

Extinction. S. M. Stanley, Scientific American Books, Inc., 1987. A nicely written and quite detailed discussion of mass extinctions and their possible causes. Contains much information on the extinctions at the Permian-Triassic and Cretaceous-Tertiary boundaries.

Glaciation

Ice Ages: Solving the Mystery. J. Imbrie and K. P. Imbrie, Enslow Publishers, Short Hills, N.J., 1979. A very readable account of the historical development of ideas about continental glaciation, beginning with Louis Agassiz and ending at the present.

Plate Tectonics

The Ocean of Truth. H. W. Menard, Princeton University Press, 1986. An interesting account of the development of the ideas of plate tectonics written by a geologist who was himself involved in many aspects of the revolution in the earth sciences, and who knew personally most of the other major figures. The textbooks listed earlier all give clear accounts of the subject of plate tectonics; this book provides much of the historical background.

Climate and Climate Change

Weather, Climate and Human Affairs. H. H. Lamb, Routledge, 1988. A collection of papers and essays by the author that detail the causes and effects of climate change, concentrating on the historical period.

Volcano Weather. H. Stommel and E. Stommel, Seven Seas Press, 1983. The subtitle for this book is *The Story of 1816, the Year without a Summer.* The authors have researched the effects of the eruption of Mount Tambora in Indonesia in 1815, and this delightful little book describes what they found in an interesting, informative way.

Climate, Man, and History. R. Claiborne, W. W. Norton & Company, 1970. A lively and well-written account of climate, the ice ages, the effect of climate on human evolution, and many other related topics.

INDEX

A SHORT HISTORY OF PLANET EARTH

For Sheila, Christopher, and Katherine

Copyright © 1996 by J. D. Macdougall
Published by John Wiley & Sons, Inc.

Library of Congress Cataloging-in-Publication Data
Macdougall, J. D.
    A short history of planet earth : mountains, mammals, fire, and
ice / J.D. Macdougall.
     . p.   cm.
    Includes bibliographical references and index.
    ISBN 0-471-14805-9 (cloth : alk. paper) /
    ISBN 0-471-19703-3 (paper : alk. paper)
    1. Historical geology.   I. Title.
QE28.3.M33   1996
551.7—dc20   95-46399